Speaking for a Long Time

ADRIENNE L. BURK

Speaking for a Long Time

Public Space and Social Memory in Vancouver

UBCPress · Vancouver · Toronto

19 18 17 16 15 14 13 12 11 10   5 4 3 2 1

Printed in Canada with vegetable-based inks on paper that is processed chlorine- and acid-free.

LIBRARY AND ARCHIVES CANADA CATALOGUING IN PUBLICATION

Burk, Adrienne L., 1954-
Speaking for a long time : public space and social memory in Vancouver / Adrienne L. Burk.

Includes bibliographical references and index.
ISBN 978-0-7748-1698-4 (bound)
ISBN 978-0-7748-1699-1 (pbk.)

1. Memorials – Social aspects – British Columbia – Downtown Eastside (Vancouver). 2. Public art – Social aspects – British Columbia – Downtown Eastside (Vancouver). 3. Public spaces – Social aspects – British Columbia – Downtown Eastside (Vancouver). 4. Collective memory – British Columbia – Vancouver. I. Title.

FC3847.5.B87 2010   971.1'33   C2009-907099-5

---

e-book ISBNs: 978-0-7748-1700-4 (pdf); 978-0-7748-5933-2 (epub)

Canadä

UBC Press gratefully acknowledges the financial support for our publishing program of the Government of Canada through the Canada Book Fund, the Canada Council for the Arts, and the British Columbia Arts Council.

This book has been published with the help of a grant from the Canadian Federation for the Humanities and Social Sciences, through the Aid to Scholarly Publications Programme, using funds provided by the Social Sciences and Humanities Research Council of Canada.

UBC Press
The University of British Columbia
2029 West Mall
Vancouver, BC V6T 1Z2
604-822-5959 / Fax: 604-822-6083
www.ubcpress.ca

Contents

Preface

What does it mean, in contemporary societies, to remember? Historically, the official narration and crafting of memory has belonged to social elites who have used all the mechanisms of public memory – rituals, images and propaganda, and the design and placement of spaces for social engagement – to engineer alignments of individuals with wider social identities. In times of panic, elites have used memory devices to resuscitate heroic pasts or sketch prosperous futures, and in times of stability, to arouse grand ambitions. But now, just as memory crafting itself is shifting under the weight of the near-infinite content afforded by ubiquitous digital technologies, so too are other social fundamentals that have traditionally been used to align individuals within larger social contexts.

For example, consider how, from the ancient village founded on sacred geometries to the megalopolis of contemporary urban sprawl, humans have always deliberately fashioned areas in which to mix, to display, and to enact humanity. Precisely because public spaces have offered this mixing of personal and social realities in plain view, the notion of the self has been largely formed and negotiated in relation to particular *public* spaces: commons, arenas, markets, parks, stadiums, plazas, squares and streets. But with revolutions in printing, transportation and telecommunications, this place-dependent relationship of identity with particular sites has shifted. Now it is unusual to find people anywhere on the earth entirely unaware of other places and peoples; the exclusively face-to-face world of knowing is nearly gone. Instead, individuals experience varying degrees of access and mobility across and varying spatial identities in many different kinds of public spaces. We interact meaningfully with people and images and ideas born in places we will never walk. The once foundational relationship between place and identity is increasingly unstable.

This is not the only fundamental shift. For decades now, a veritable parade of identities, whether those of nationalisms, diasporas, gender, race, sexuality, (dis)ability or political affiliations, has problematized our understanding of what, if anything, belongs to all, and how. Even the memories of successful collective enterprises (such as the origins of Canadian public health care; the truth commissions of Chile, East Timor and South Africa; the US Civil Rights movement; or the election of Barack Obama) are conventionally summarized in anecdotes of a few extraordinary or charismatic individuals. The more such cultural amnesia about collective action takes hold, the less we take for granted the very idea that individuals are part of a public and the more we hasten the day heralded by Margaret Thatcher's comment, "There is no such thing as society, only individuals."

Even religious affiliations, a possible source of societal integration, have become in recent years an increasingly tense array of strident fundamentalisms and secularisms. Taken together, these foundations for constructing identities – one between place and identity, others designed to help us construct a sense of self within historical and/or moral contexts – appear to have given way. Now, in a world mediated more by streams of decontextualized images, sound-bites, and 24/7 information than by complex discussion and nuanced rituals, it is difficult to construct a contemplative sense of self, and we have also lost opportunities for honing the skills and attitudes necessary for working productively with others.

The question of what to valorize and how to meaningfully remember in these shifting circumstances is provocative. Especially as our political leaders increasingly make public apologies for genocide, or apartheid or internment as a first step in cross-generational reconciliation, and especially as the truly global realities of climate change, food security and financial interconnections grip the peoples of the planet, it becomes ever more pressing to understand how it is that we actually make the societies in which we live. How do we decide, and mark, what is possible, what is warned against, what is dreamt? How do we use memory meaningfully?

As a geographer, I found myself drawn to examine this question in an empirical way, by contemplating the built spaces of a modern Western city. But I should clarify: critical human geographers theorize spaces as existing not only physically, but imaginatively and representationally as well. And so, when in the spring of 1996 I encountered an unusual civic stalemate in Vancouver, British Columbia, I found a case that suggested rich analytical promise. The stalemate was this: a monument had been conceptualized, funded, designed, argued about and even sited in city plans, yet remained unbuilt. I was intrigued with the idea of a monument existing robustly in the imaginative and representational domains, but blocked from the physical domain.

I began my research and then was startled when, after seven years of delay, there came surprising news: groundbreaking for the monument was imminent. Within days I found out about another monument (just built), and yet another (to be built within a few months), all within walking distance of the first one, and all addressing quite similar issues – violence, in particular violence against women. Even more remarkably, it became apparent that all three monuments had been supported by the collective efforts of the socially stigmatized. Thus, people who were among the least likely to be able to set the terms of a social encounter had nevertheless attempted, and succeeded (in three separate instances), in one of the most difficult acts of cultural marking – to install, with permanence and in full view, a monument that acknowledged disturbing social facts.

Creating markers is not in itself an unusual activity. All types of markers, from the inuksuit of the Arctic to the stone cairns of the Maori, suggest that as a species, humans like to signal each other and telegraph information: "choose this way," "one died here," "this place is sacred." It has been a human impulse to adorn the dramas of life's passages with such forms. But to make monuments, in this day and age and in one of the most expensive and contested areas of real estate on the continent, is a remarkable enterprise, given how logistically, politically and financially difficult they are to install.

And so to a brief introduction of the monuments themselves, which are in Vancouver's Downtown Eastside: *Marker of Change*, the CRAB

Park boulder, and *Standing with Courage, Strength and Pride.* Despite what the monuments appear to have in common, their differences are also intriguing. Only one is perceived (and then not uniformly) as *of* the neighbourhood; the others feature in quite contested negotiations of ownership. Who is welcomed, and how they are welcomed, in the spaces surrounding the monuments is extraordinarily nuanced. How the monuments are used in the imagery of the neighbourhood, and in the imagery used to address the issue of violence, both for rituals and for memory-making, varies widely. The monuments have all been well received, as attested by the fact they are regularly used, tended and not defaced, but they are not sited in a typical neighbourhood for monuments (nearly all of the dozens of Vancouver's other civic monuments are housed elsewhere).

Vancouver's Downtown Eastside is a contemporary neighbourhood that sits on lands that have been intensely used and inhabited since they emerged 11,000 years ago from glacial retreat. The location's nearness to water influences not only human traffic, but also plant and animal life. These lands have served as destination sites, places of abundance and places of collective social activity – whether as First Nations villages or as rallying points for labour strikes. In contemporary terms, the Downtown Eastside physically includes several neighbourhoods, as well as the streets linking them: Chinatown, Gastown, Strathcona, Victory Square and the Hastings Street Corridor, a total area measuring approximately three kilometres east-west, another two north-south. For those who live in the Downtown Eastside, it is an area known almost entirely by foot. For those who do not, it is most often known as an area one passes through, situated as it is near the downtown business core and bisected by bus connections and wide cross-town traffic corridors.

There are many contradictions in this neighbourhood. Though developers who eye the prime real estate it occupies (it sits directly next to the central business district) try to portray the residents as transient, in fact the Downtown Eastside has until recently been the second most stable neighbourhood in Vancouver, after the elite enclave of Shaughnessy. It is full of artists and people deeply loyal to that area. Traditionally populated (since European contact) by seasonal labourers and the low-waged or unemployed, some contend that the Downtown Eastside residents have paid rent so regularly and in such amounts over decades that they are entitled to a claim of property (Blomley 2004). But for many decades now, the neighbourhood has been hard hit by numerous changes. In the words of local community historians:

> Vancouver's origins are based in [the Downtown Eastside]. Gastown's sawmills helped to spawn a commercial zone along Hastings Street. Eventually, head offices, banks, theatres, hotels and department stores all set up shop there. It was home to the main library ... and City Hall. Hastings Street was also a key transportation hub – a streetcar terminus located at the B.C. Electric Building allowed riders to catch connections to other parts of the city.
>
> In 1958, streetcars stopped running in the area, taking away the daily infusion of pedestrians ... [soon after] many head offices began to follow suit. As a result, traffic to the neighbourhood decreased by almost 10,000 people a day. The gradual loss of low-income housing in other parts of town ... drove more people to the increasingly affordable Downtown Eastside ... by the 1970s, lack of public funding led to the de-institutionalization of thousands of psychiatric patients, who found the Downtown Eastside to be their only affordable housing option.[1]

Demographically, in the most recent figures, the neighbourhood is home to approximately 16,500 people, 67 percent of whom are low-income, two-thirds of whom are men, and as noted, many of whom suffer from mental health disorders.[2] Further, though the Downtown Eastside houses people from around the world, it is widely acknowledged within Canada as having one of the largest off-reserve concentrations of Aboriginal peoples of various First Nations. But one of the most difficult body blows to the neighbourhood began in the late 1980s, when cocaine became widely available, resulting in a downward spiral of negative social consequences. Now, it is possible to qualify as a senior citizen at the age of 45 within the Downtown Eastside because life expectancy is so low. In recognition of the systemic violence, particularly against Aboriginal

1 http://vancouver.ca/commsvcs/planning/dtes/communityhistory.htm.
2 http://vancouver.ca/commsvcs/planning/dtes/pdf/2006MR.pdf.

women of the neighbourhood, there is an annual march through the streets every St. Valentine's Day.

Given the heritage, multicultural, health, crime and property complexities of the Downtown Eastside, in March 2000, the City of Vancouver, the British Columbia government and the Canadian government signed the Vancouver Agreement, a five-year commitment to support sustainable community and social and economic development in the area. This has resulted in another contradiction: though the Downtown Eastside is one of Canada's poorest neighbourhoods, it is also awash with money. Four and a half million dollars came from the Vancouver Agreement alone, but millions more are set aside every year for scores of social services that are clustered in the few blocks of the area. One Canadian Press wire article claimed this amount neared two hundred million dollars/year (cited in Woolford 2001).

In this neighbourhood, the three monuments were installed between July 1997 and June 1998. However, these dates are slightly misleading without context. Though the first monument *(Marker of Change)* was conceptualized in 1990, it took seven years to realize, not least because of the tremendous controversy it generated. The second (the CRAB Park boulder), by way of contrast, was imagined and realized very quickly. And the third *(Standing with Courage, Strength and Pride)* was an idea that nearly came together at least once before it was actually undertaken; work on it began on Labour Day, 1997. Thus, a more accurate way to read the dates of when these monuments became material would place them as follows: the CRAB Park boulder, July 1997; *Marker of Change,* August 1997; *Standing with Courage, Strength and Pride,* September 1997.

In order to tell the stories of these three monuments, and to link them with wider debates, the book follows in three parts. In Part 1, each of the monuments is presented in some detail, drawing on research gathered over four years, including archival research, more than 80 semi-structured and naturalistic interviews with the artists and advocates for each monument, participant observations in neighbourhood meetings and events and analysis of key national, municipal and neighbourhood film and print media, as well as more than six years of neighbourhood artwalks, marches, and encounters at the monument sites themselves. Readers will thus find interspersed in the text archival materials, interview excerpts,

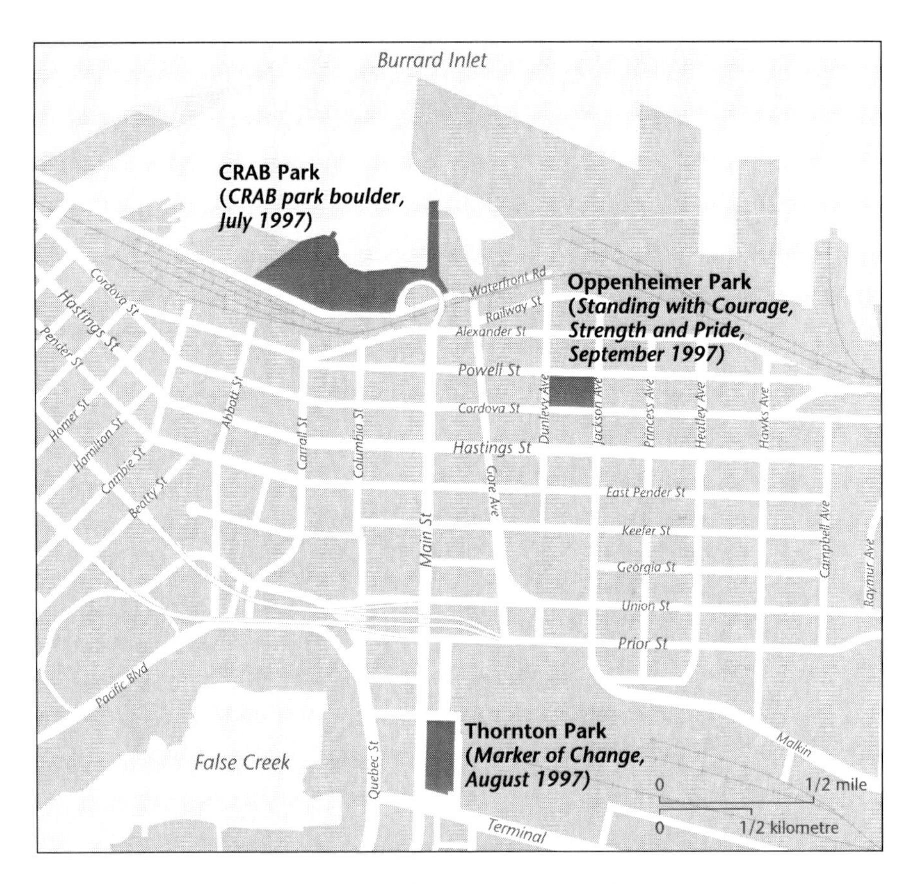

Part of what is so unusual about these three monuments is that they were erected only blocks away from each other and within three years. | *Cartographer: Eric Leinberger*

field-note observations and quotations from published material. The intent in Part 1 is to introduce the origin stories of each monument, so as to highlight the perspectives of its advocates and to introduce its design features, placement, wider social contexts and current usages. It is important to note that the relative weighting of the accounts in Part 1 is unequal. This is due to several factors. Most notably, more advocates were alive and thus available for interview for *Marker of Change,* and the archival sources for the three monuments varied dramatically, from the abundant (for *Marker of Change*) to the nearly non-existent (for the boulder). Also, as will be more fully explained, advocates and constituencies felt very strongly about the role of the media in regards to the monuments, and in many cases banned photography of key events. I have tried to address this apparent inequality of treatment through the inclusion of other related images and texts of the Downtown Eastside, but I wish to stress that the weightings here do not correlate with the relative importance of the monuments in the social fabric of the neighbourhood, as I expect will become clear in the pages that follow.

Part 2 introduces a broader and more theoretically informed discussion about public space, social memory and monuments, with a particular emphasis on how these intersect within a geographic sensibility. Part 3 returns to the details of the narratives in Part 1 in light of the theoretical discussions of Part 2, and articulates the constellation of factors that permitted such extraordinary and unlikely monuments to manifest. These factors include not only those elements we would think of as essential to effective community organizing, but also a profound engagement with memory, space and place. This section also speaks to what lessons the rest of us can take from these monuments.

For these stories provoke us to discover and create such images within our own communities: as these cases reveal, the difficulty of creating such symbols does not excuse us from taking up the challenge. Championed by the poor, the traumatized, the discarded and the marginalized, the monuments examined here could hardly have had less auspicious beginnings. Their organizers began their efforts without offices, money, public knowledge or sanction. They endured the indifference and even hostility of powerful forces in the form of the media, the police, civic institutions and individuals. Normally in North American individualism-obsessed cultures we ascribe such achievements to the presence of highly charismatic persons with great social capital. But such an analysis diminishes us all. Things are possible collectively that are not possible individually, as these monuments attest. These activities require skills that we do not often see highlighted: those that nurture cooperation, tolerance, trust and collective problem-solving. Today each of these three monuments stands, and is used, honoured, and continuously engaged in civic life. Their installations and ongoing presence contribute, without doubt, to changes on multiple levels: the image of violence as a public rather than private issue, the political landscapes of Canada's legal and police systems, Vancouver's civic elections, the city's imagined civic identity in the grip of the 2010 Olympics and, specifically, the City's responsibility and accountability to the residents of the Downtown Eastside. I offer this book in acknowledgment and appreciation of these stunning accomplishments to all those who inspired these monuments. May this book make the circle of attention and healing to those *you* honour that much larger, and in turn illuminate for us all what it can mean to actively remember.

Acknowledgments

I am grateful to so many for examples of what emerged as the key qualities in this work: how to listen across difference, how to be brave enough to live with confusion and how to trust in human decency. I appreciate most of all the many decades of conversations and experiences exploring these things with my family: thank you, Tom, Bob, Pat, Elliott and Darcie for all you teach me, by design as well as by how you live your own lives. For similar help, and also for their thoughtful guidance with the original research, I particularly want to thank Ellen Gee, Dara Culhane, Nick Blomley, Michael Hayes, Sue Ruddick and Bev Pitman. Some periods during this project were especially challenging. For those of you who saw me through to the other side of them, my thanks to Jennifer Hyndman, Laura Carlson, Budd Hall, Sumiko Nishizawa, Lorraine Gibson and Sean Markey. For friendship and support throughout this project and many others, thanks also to Aurian Haller, Damian Collins, Tara Fenwick, Janice Bearg, Chris Gilmour, John LaBrie, Marti Roach, Pat Nicholson, Janet and Anne Ericsson, Ed Taylor, Ed and Lucille Broadbent, Jack O'Dell and Jane Power, and the estimable Harley.

I wish to thank as well my UBC Press editors Darcy Cullen, Ann Macklem and Jillian Shoichet, and two generous and unknown readers, all of whom helped me with many aspects of preparing this manuscript. Jean Wilson also gave early and much needed encouragement. Tara Adair and Greg Ehlers have been beyond wonderful to work with throughout. In addition, I would like to thank Sandy Cameron and Bud Osborn for giving me permission to include their strong poems; I much appreciate this thoughtful and gracious generosity. I am also most grateful to those who helped me with archival work at both Simon Fraser University and the City of Vancouver, and to the Social Sciences and Humanities Research Council, who funded the original research.

Finally, to Beth Alber, Fred Arrance, Dick Baker Sr., Kim Bruce, Janine Carscadden, Steve Johnson, Sharon Kravitz, Don Larson, Julia M., Krista Marshall, Christine McDowell, Vanessa Pasqualetto, Lianne Payne, Lindsay Seltzer, Moira Simpson, Dan Tetrault, Elinor Warkentin and the participants in the Community Directions meetings – thanks for your generous and thoughtful participation, and for permission to build from the stories you shared with me.

Speaking for a
Long Time

PART ONE
ACT

Marker of Change/À l'aube du changement

The origins of the first monument can be precisely dated: December 6, 1989. That evening, in Montréal, Québec, a male gunman entered a university classroom, methodically separated the women from the men, and ordered the 51 men to leave. To the women who remained in the room, he said, "You're women. You're going to be engineers. You're all a bunch of feminists. I hate feminists." Nathalie Provost replied, "Look, we are just women studying engineering." He shot her in the leg. He shot all the other eight women in the room too, then went floor to floor, shooting in all twenty-three women and four men. Twenty minutes later, thirteen female students and one female office worker were dead, and another thirteen were left injured. The gunman then killed himself. The suicide note in his pocket was quite explicit: "Feminists have ruined my life." The third page of the note was a hit list of 19 prominent women, including union leaders, journalists and 6 female police officers who had beaten their male colleagues in a game of volleyball a few weeks earlier (Fitterman 1999).

What became known in Canada as the Montréal Massacre was, indisputably, a hate crime. It was directed against women in a university engineering school precisely because they were women in a university engineering school, a professional training environment that the killer felt was inappropriate for women. Canadian reaction was unprecedented: a nationally televised state funeral in Montréal, thousands of minutes of media coverage in every medium and much public soul searching. As the coverage began to focus increasingly on the notoriety

It goes on one at a time,
it starts when you care
to act, it starts when you do
it again and they said no,
it starts when you say *We*
and know you who you mean, and each
day you mean one more.

— Marge Piercy, excerpt from *The Low Road*

of the murderer, and as the female victims' names faded from public memory, a group of very determined women in Vancouver set about to make a national monument.

Though the idea of the monument began as a spontaneous response to the Massacre by a single student, crucial institutional support was given early on by Capilano College in North Vancouver. Although members of the original monument project group indicated to me that probably over the years some 40 women were significantly involved, there were 8 who oversaw the final realization of the monument. Like the original group and most of its members over the years, these eight were all working women of European-Canadian background (though of different national traditions). Many had spent years in, or were currently still working in, front-line services (for example, rape crisis centres, battered women's counselling and related services). They ranged in age from their early twenties to near retirement. What remained consistent from early on until the day of the monument's installation was a commitment to name the murdered women and leave the murderer's name unspoken. The murdered women's names are Geneviève Bergeron, Hélène Colgan, Nathalie Croteau, Barbara Daigneault, Anne-Marie Edward, Maud Haviernick, Maryse Laganière, Maryse Leclair, Anne-Marie Lemay, Sonia Pelletier, Michèle Richard, Annie St-Arneault, Annie Turcotte and Barbara Klucznik Widajewicz.

Support for a monument was widespread but not unproblematic. Not only were there the considerable (and in the beginning, unforeseen) complications of actually commissioning such an object, but there was also ambivalence about using the monument form. One of the organizers explained: "Feminists particularly criticized the idea of a monument, I think. Quite effectively ... and had quite sophisticated arguments against it ... monuments had a bad rep. Because, I think initially, all the monuments around the World Wars were out of grief. But they don't look like that to later generations ... I think that they look bad. They look like they're perpetuating war ... and that they are glorifying war, and so monument making ... monuments were intrinsically a bad way to go." Yet the idea persisted in part because of the personal experiences and knowledge of the early organizers. One advocate recounted how the idea of a monument took hold for her:

> You know, I was in media ... I knew that I needed to do work on violence against women, and I wanted to do something that wasn't front-line rape crisis. I knew I couldn't do that any more ... And then ... ironically the massacre happened my first semester, and I didn't want to do any of the things I was learning. I felt that permanency was a true answer to what happened ... I already knew that millions of women had been murdered, just during the times when the so-called witches were being massacred over those centuries, and I knew that the day-to-day murder had been happening for thousands of years ... of women and children. And I knew about rape and battery, and I'd been on a rape crisis line, and I knew that women were so oppressed that there we were only starting to acknowledge violence against women, really, collectively, and that it hadn't been going on that long ...
>
> [I was also discovering while] I was at school ... that I had worried so deeply about my sisters. Because not only was I abused, but I had three younger sisters. And that it had just been this torture for me. Like, the ramifications of more than one victim, and, and loving the other people involved, and, and not being able to protect them. I was the oldest daughter and, and I couldn't protect them. I couldn't. And it was just this horrible thing that I carry with me ... And this really motivated me. I felt that men use permanency very effectively. You know, we've seen them perpetuate real evil through permanency. But we've also seen good men try and do something

> with permanency that's constructive, in terms of memory ... In terms of remembering bad things, so that they don't happen again ... I really felt that women needed to be memorialized ... There's a whole level of discussion where permanency ... if men use it, it's got to be bad. If it's part of patriarchy, it's got to be bad. But I think that's not true. I thought that there was a way to redeem permanency and use permanency in a constructive way.

As the organizers were to discover, however, the creation of a monument, especially one that involves a national submission and jury process, was an enormously expensive proposition, both in money and in time.

First, beginning in the spring of 1990, a series of meetings was held at Capilano College, now a university college in the city of North Vancouver, where one of the monument's advocates was enrolled. College faculty, students and staff came together, and the women's centre on campus donated office space and equipment. The idea of a monument was present from the beginning, but national reactions to the events in Montréal were still unfolding. For example, a woman who had been in the building in Montréal but who had not been shot undertook to co-ordinate a memorial that was sensitive to the school and families. Later, she devoted herself to a (largely successful) multi-year campaign to address gun legislation. A filmmaker decided to analyze the media responses to the event so as to underscore how certain voices of authority, such as the CBC, tried to gloss the event as about a "crazed madman" rather than as a crime specifically against women. The New Westminster MP, Dawn Black, began drafting a private member's bill that in 1991 would declare December 6 to be a national day of remembrance, education and action around violence against women. The Capilano College group, though it retained a strong organizing focus, was itself a remarkably fluid entity until 1993:

> Anybody could come and go. And that's how we wanted it. At first, you know, you have to do it that way ... And ... we'd have meetings where you'd be completely undone. Like, everything would be in question. "Well, why even do a monument?" And I mean, you're three years into your project. So there [were] no

> reference points to even discuss what was still just an idea, which was the idea of a competition, and the idea of a monument ... For every woman who came in the door and sat ... whether it was 3 meetings or whether it was 30 or 300 meetings. Everybody had to go through their own process. And we allowed that. It, it was very time-consuming.

Initially, the Capilano College student who conceived of this project, Christine (Chris) McDowell, believed the monument should be located in Montréal. However, various factors combined to challenge that thinking. First, the tragedy was still raw in Montréal, making it harder for a project such as this to come together so quickly, a difficulty that some of the parents of the murdered women spoke about eloquently. In addition, many Montréal activists found themselves facing issues of security and the palpable public fear that a similar massacre might happen again. At the same time, Capilano College support for the project was strong. Thus, the decision was made by the advocates to establish a monument in a Vancouver public park. This initiated a cascade of considerations. As Chris McDowell said, "We knew we were breaking the rules. We knew there would be consequences. We just didn't know what they would be."

The Vancouver Board of Parks and Recreation (the Park Board) made it clear that since the petitioners didn't constitute a "neighbourhood" or "recognized community," the committee would first have to demonstrate that it did represent a community. This was a slightly surrealistic undertaking: in the national conversation about violence against women, there was only really one side to take; yet despite this, the City of Vancouver still wanted the committee to demonstrate that it spoke for a community of interest. In any event, the petitioners gathered a series of letters from a wide range of people in favour of the project – from MPs to therapists to front-line service workers (see text box on p. 6).

When the Park Board was satisfied that there was sufficient support for the monument project, deliberations about the site itself began. Criteria were developed by the Monument Project Committee and by the Vancouver Park Board and the Vancouver Public Art Committee, as well as the Downtown Eastside community associations and other

Women's Counselling Services of Vancouver
Audrey McLaughlin, MP, NDP leader
Hon. Mary Collins, MP, Minister Status of Women
Battered Women's Support Services
Status of Women Committee of the College Institute Educators Association
Department of Women's Studies, Simon Fraser University
Svend Robinson, MP, Burnaby-Kingsway
South Surrey/White Rock Women's Place
Darlene Marzari, MLA, BC Minister Tourism and Culture
Senator Pat Carney
Rosemary Brown, former BC MLA
Libby Davies, Vancouver City Councillor
James Cormack, Counsellor
Joy McPhail, MLA, Vancouver-Hastings
BC NDP Women's Rights Committee
Margaret Birrell, Director BC Coalition of People with Disabilities
Margaret Mitchell, MP, Vancouver East
Sheila Copps, MP, Hamilton East
Linda Reid, MLA, Richmond East
Dawn Black, MP, New Westminster-Burnaby
Metro Action Committee on Public Violence Against Women

— from a press kit, "What They're Saying about the Women's Monument Project"

stakeholders. These criteria included "accessibility, safety and security issues, ambience and permanency."[1] More than 40 sites were considered and 24 were thoroughly evaluated (see text box on p. 7). There was a strong interest by the Monument Project Committee in the new Science World park, and local community associations expressed support for this location; however, the Strathcona Community Association recommended Thornton Park as preferable because of its working-class roots and accessibility to working people.

Today, Thornton Park is 3.8 acres of Edwardian-style plantings and pathways laid out on the doorstep of the Canadian National Railroad terminus, offering a sort of gateway glimpse of Vancouver. Indeed, it was designed as a city park when Vancouver was chosen over previous favourite Prince Rupert to become the west coast destination for the

1 Simon Fraser University Archives, Women's Monument Fonds, F-101-5-0-10, "Why Thornton Park?" Part of a package referred to as "Sept. 30, 1997 Summary of Project Major Events, People, Aspects and Issues, handed out to media."

Marker of Change sites considered (as of January 9, 1993)[2]

BC Hydro Plaza – hydro executive contacted WMP "shelter good for rallies"
Queen Elizabeth Park – no washroom, convenience, food, lighting, transit and telephone
Nat Bailey Stadium – open, plain, poor access to transit, phone or seating
Nelson Park (Thurlow, Nelson, Comox)
Barclay Heritage Square (Barclay and Broughton)
English Bay, Sunset Beach – overused, area is swamped, several pieces of art
Alexandra Park (Bidwell, Burnaby, Beach), English Bay Beach
Charleston Park (near False Creek 2nd Ave.)
Sutcliffe Park (near False Creek 2nd Ave.)
Granville Island (where?)
UBC, Capilano College – INACCESSIBLE, CANCELLED
Plaza of Nations
EXPO site in general (where?)
Trout Lake (John Hendry Park)
Thornton Park
Strathcona Park (Venables between Main and Clark) – very dangerous to women, limited green park space in area – CANCELLED
new park next to Science World (Creekside) – science theme ties in women, engineering, science
Pan Pacific Plaza
Plaza at Hastings and Hornby, North Side
Plaza at 666 Burrard (behind Christ Church Cathedral)
Robson Square – CANCELLED
new library – library and director "no controversial art here," not ready until '95 a problem
Langara – too restrictive for access, convenience, parking, pedestrian
Grandview Park – CANCELLED, max. space being used

national railroad. But the park's earlier incarnations are revealing. For some 3,000 years the area was an intensely fertile wetland, a one-mile tidal fish trap of sole, perch, sturgeon, flounder, perch and smelt. A trail wound from the south Fraser River (today's New Westminster) through forests dense with fir, hemlock, crabapple and thousand-year-old cedars, all the way to Burrard Inlet. The clearing that would later become the park was a stark contrast: "The land is low, covered with second growth,

2 Simon Fraser University Archives, Women's Monument Fonds, F-101-6, "Site Choices."

Marker of Change | Note the understated way the monument sits alongside other benches | *Photo: Greg Ehlers*

old stumps, wet and swampy, sometimes flooded at high tide. In season there were flocks of water fowl, mallards, butterballs, herons, loons, all very wild, on False Creek waters. The Squamish lived at their village at Snauq and sometimes passed up the creek, and, at high tide only, paddled through to Burrard Inlet – about Campbell Avenue."[3]

The Squamish people referred to the land here as "Khiwah'esks" (two points exactly opposite) because for thousands of years the land nearly touched over False Creek at the site of the present-day rail station; beyond, where the rail yards now lie (and near where the 2010 Olympics athletes' village is built), lay a large lagoon. The first Europeans to the area built bridge after bridge over the waters at this point, each bridge collapsing in turn. In the 1880s, when it was decided that Vancouver would be the end of the rail line, the CPR and the City of Vancouver began to reshape the area in earnest. By 1918, the Main Street hill had been levelled three times, and the more than 27 acres of lagoon had been filled with 20,000 cubic yards of soil brought from Chilliwack (100 kilometres east of Vancouver), a feat engineered and financed by Henry Thornton, then General Manager of Canadian National Railways. The Park Board Annual Report of 1925 called the development "one of the most important under-

3 City of Vancouver Archives, DIST pg. 93 N91, caption on 1898 photograph.

takings of recent years ... Within a year from the commencement of operations the whole area was transformed into a park of lawns, flower-beds, trees and shrubs, with cement walks and an ornamental lighting system. An expenditure of $30,000.00 was involved."[4]

In 1923, the park was formally dedicated and named after Thornton. But in 1928 the Canadian National Railroad and the BC government swapped land, and the ownership of Thornton Park became legally and imaginatively ambiguous. In 1934, the City voted it the most desirable site for the new city hall, but for unexplained reasons, the city hall was built instead at its present location, about 2 kilometres south. When Thornton died, a plaque honouring him was mounted inside the rail station by the locals of 17 brotherhoods of organized labour. The boosters still preferred to focus on Thornton Park's aesthetics, comparing them favourably with those across the continent.

Organized labour continued its presence in Thornton Park by using it as a site for rallies in the 1940s, decorating a 22-foot-high pylon with murals against fascism. In 1947, a long correspondence began, citing the original purchase agreements from 1924 about the proportions of support for maintenance that were to be borne by the Park Board and the railroad. Except for this episodic correspondence, which details diminishing interest by both the Canadian National Railroad and the City in maintaining the landscaping, the official City of Vancouver Archives has no record of activity about the park during the remainder of the 1940s, 1950s and 1960s, and much of the 1970s, save for a notice of an anti–Vietnam War rally held there.

During these years, however, much was changing in the neighbourhood around Thornton Park. Vancouver is a relatively new city, having incorporated in 1886. Like many cities, it was largely founded on the systematic dispossession of the indigenous communities and re-populated by migrant workers who immigrated or were brought in to build a frontier resource economy. Not inconsequentially, these same residents proved tenacious over the following decades in resisting their own displacement: "[the area's] proximity to the industrial waterfront and the large numbers of industrial workers who lived in and moved through the neighbourhood

4 City of Vancouver Archives, PDS89, Thornton Park, Park Board Annual Report 1925, 27.

"No railway station entrance in the United States or Canada could compete with Thornton Park, the beauty spot in front of the station."
— "Park Praised," 1940[5]

meant that it was a key site of labour militance for much of the century. The supportive community in which this kind of oppositional culture thrived provided a vital legacy for the mobilization to improve living conditions that began in the late 1960s and early 1970s" (Sommers 2002).

Two key events fostered the activism of the 1960s and 1970s. Via Herculean efforts spearheaded by the Downtown Eastside Residents Association to redefine the area in popular perception as an actual *community* rather than a skid row site of transients, and through a coalition of community groups and individuals that successfully rejected a freeway construction that would have completely reconfigured the city centre, the area began to improve in terms of housing, health, parks and recreation. In other respects, however, this part of the city was getting markedly more dangerous. Beginning in the late 1970s and early 1980s, sex trade workers, who were often Aboriginal and often addicts living and working in the Downtown Eastside, were found murdered or were reported "missing."

Moreover, the frenzy created around Expo '86 targeted precisely this area of the city, disrupting it further. More than a thousand residents were evicted from their homes in the Downtown Eastside. Thornton Park was developed as a presentation space for the city: its bordering streets were widened, and both Science World and the Main Street SkyTrain Station were built. By the year of the Montréal Massacre, Thornton Park was back in the City's sightlines as an underused public space. When the idea of the monument coalesced, one of the organizers recalled a Park Board commissioner saying: "[it] would be good to have a memorial in that park ... [as] a neglected park [the monument] would create a focal point ... the park ... wasn't being used. It was unsafe for people. And it was in ... a space that was ... it's sort of rundown hotels around it, and

5 City of Vancouver Archives, MSS54, Vol. 19, Thornton Park, unidentified newspaper press clipping, "Park Praised," May 11, 1940.

people never used it. And if we had the monument there, it would ... make the space more public and useable."

As mentioned above, more than 40 sites were considered for the monument over a period of a year, and the manner of those considerations was quite inventive, including visits to the sites at all times of day and night, picnics and rituals at the sites and observing who came and went through the spaces. There was an interest from the beginning in high visibility for the monument and in completing its installation rather than having it lost in years of planning because a site was particularly contentious. Of course, the organizers also wanted to include rather than antagonize the public and realized that the distribution of parks (and therefore of public play space for children) was markedly lower in Vancouver's Downtown Eastside than in other areas of the city. The chair of the Public Art Committee of the Park Board, Bryan Newson, suggested that two constituencies needed to approve a possible location. The organizers took his advice, and noted in their minutes of August 25, 1992, to "convince a *broad* and a *local* community. For example, [for] the two parks around the Main St. SkyTrain Station there's little to worry about in terms of an immediate local community, but lots in a larger context (but he doesn't know how to get that response)."

In the end, the committee selected Thornton Park because it was accessible and fairly well lit, it was equipped with phones and there were workplaces nearby (should a woman need help), and it was unlikely to be sold for development due to its symbolic value and, hence, possible protection as a heritage park. A further consideration was to prove profoundly important: "positioning the Monument near the Downtown Eastside – where many women are murdered – will help draw much needed attention to this reality."[6]

Once the site was chosen, it was possible for the committee to put together the design criteria and start the process of sending out notices for the national competition. There were two certainties and one near-certainty at this point: the inscription, the necessity of including the

6 Simon Fraser University Archives, Women's Monument Fonds, F-101-5-0-10, "Why Thornton Park?" Part of a package referred to as "Sept. 30, 1997 Summary of Project Major Events, People, Aspects and Issues, handed out to media."

14 women's names, and the site. The Park Board reserved for itself an "out" as far as the monument's location was concerned. The arts consultant for the board's recreation department, Susan Gordon, noted, "the planned pathways, the plaque, and the as-yet-undetermined size of the structure make it hard to determine if [the monument] adheres to park board guidelines for permission to situate on public land ... 'I don't necessarily see it as being simple. I don't foresee instant approval'" (Dunphy 1992).

Indeed, the committee was already running headlong into some harsh resistance that it had not anticipated – resistance from those it had imagined might be supportive. Given the realities of the need for front-line services and their pathetically low levels of funding, a tremendous amount of discussion was generated about the value of campaigning and organizing for a monument rather than, for example, fundraising to aid direct services. One of the organizers, Krista, recalled: "People were saying, 'Don't you get it? Like, you have no idea what it's like, what the need is.'" One advocate mused that such stances were particularly ironic, given that it was precisely *because* so many of the committee members had worked or were currently working within direct services that they felt drawn to the originality, "hope and proactivity" of the monument project.

One of the most articulate and painful letters the committee received came from Vancouver's Women Against Violence Against Women (WAVAW), a rape crisis centre that had sponsored the December 6, 1991, vigil at which it had invited committee representatives to speak. This letter, which arrived six months after the vigil, is two single-spaced pages; it is only excerpted here. First, the letter notes the "error" of using a monument as a form: "We feel that in proposing to build this monument to the 14 women murdered at L'Ecole Polytechnique, you have succeeded in perpetuating the myths of the mainstream media, by romanticizing their deaths and setting up these women as martyrs. To erect a monument is, in our opinion, to buy into patriarchal models of glorification of the dead. We do not believe that such endeavours do anything to further feminist goals of ending violence against women."[7]

7 Simon Fraser University Archives, Women's Monument Fonds, F-101-2-0-1, letter received from WAVAW, June 10, 1992.

Second, there is the claim about money:

> We understand that your committee has set a goal of $300,000 for erecting this monument. This is sufficient money to maintain a paid staff of 5.25 women and provide a year's worth of rape crisis services to Greater Vancouver women who have been sexually assaulted, including a 24 hour crisis line, support groups, one-to-one counselling and [accompanying] services to police, court and hospital. To spend this money on a concrete block seems ludicrous to us ... As women we will not soon forget the 14 women ... nor will we ignore the countless other women who were murdered, abused and assaulted by men on December 6th or any other day. We do not need a monument for this. Our everyday lives as women serve as reminder enough of the threat of male violence ... we do not believe that the work you are currently doing is feminist. We suggest that you reconsider your goals and your means to achieving these goals. In spite of having invited you to speak at our Vigil, WAVAW/RCC does not support the work you are doing ... We ask that any funds collected by your committee at the December 6th, 1991 Vigil be donated to WAVAW and/or returned to the donors."[8]

WAVAW was invited to meet face to face with the committee to discuss points of difference, but the organization never responded; neither were the donations collected by the Monument Project Committee returned to the donors or donated to WAVAW. It was extremely painful, and consistent throughout all the years of the project's fundraising, that, as Janine, one of the members of the monument committee, put it, a scarcity mentality appeared to dominate "around resources. And what's available. And that somehow ... there's a small bit of money out there to, to help us survive, and, and ... we just said, 'No. We don't. We don't buy that.' There are billions and trillions of dollars out there ... for guys to support their sporting events ... You know. I mean, it's bullshit."

But in a way, the monument advocates did have a sense of playing on a larger stage: the women had a keen appreciation for the fact that the

8 Simon Fraser University Archives, Women's Monument Fonds, F-101-2-0-1, letter received from WAVAW, June 10, 1992.

spectacular nature of the Montréal Massacre and its national coverage had the potential to open doors to fundraising and awareness. From the inception of the project, all fundraising events were intentionally co-organized with other, direct-service agencies so as to support both the monument and front-line organizations. There was also an intentional strategy, because the project was art-based, *not* to compete for direct-service funding. The fundraising for the project was always envisaged as coming from donors who were giving "to the issue of violence against women for the first time;" such donations would include artistic grants from the Canada Council and funds from architectural and engineering companies.

Two other members of the committee commented that what was most troubling about the presumption of funding scarcity was how it was used to obscure both the issue of violence against women and the possibility of an alternative approach:

> Part of the, part of the pain of doing this project, when we were criticized, is that people didn't want to see that we were ... all working class or student ... class, or welfare class, or ... you know, people earning under 20 grand, mostly ... a year. They didn't want to see where it was really coming from, this thing. That it was coming from working, or student, or non-working women. That it was coming from a complete grass-roots thing. That every single cent had to be fundraised. That it was being fundraised by people who had no connections to power and money. Period. And so the wonder of it ... was always not seen. Just like the women themselves weren't seen. They were always called middle class, white privileged women who happened to get shot, or something. And again, that wasn't true who they were, either. About three came from privileged families, out of the fourteen, as far as I can tell from talking to the families. So it, it just doesn't make sense to me. What happened to us. And yet, that was how we were dismissed. And how we were objectified, and kind of nullified by people who were against it. But then, there's all these people who've perceived something closer to the truth, and responded extremely openheartedly. You know our

> letters, are, like this [hands wide apart] ... like, they're not in the middle usually, right? And the phone calls were the same way. So it was ... It was very strange. But it just shows you part of how people dismiss each other. And objectify each other. And part of it is a class thing. And in our case, it was to call us upper class, and 'screw you.' Or to see us as upper middle class, and ... 'screw you.'

Even years later, when committee members went out to ask for support or to provide education about the monument, they were frequently confronted by arguments that inaccurately cast the project as against direct services. Once they went to a local community college with a presentation tailored specifically for a college audience. In the end, the self-identified feminists on the college committee turned down their donation request, giving the $500 requested by the monument group to a front-line service instead, while at the same time criticizing the monument project as "not grass-roots" enough. Committee representatives told me that although they were very frustrated by this turn of events, "we applauded them for donating the money, because you know that was the first time they'd done that. [But I thought] You know, you sure are high and mighty for doing the right thing, but why don't you do that every year?" One member reflected that such exchanges "were not about the truth ... The point was discrediting [the project]. It wasn't about truth. And that was what's so painful ... that's when you realize you're really, you know, you're in a war of words that's not about truth between you and that person. But, but then you use it to try and educate other people because you do care about the truth."

In November 1992, the Monument Project Committee and the Battered Women's Support Services (BWSS) co-sponsored a fundraising event, *Voices for Change,* with half the proceeds going to BWSS. In addition, all records from the event, including the names, addresses and phone numbers of the ticket purchasers, were made available to the BWSS. The committee had, since its inception, claimed it would do its fundraising in this way, to enlarge the pool of funds, increase education about and deepen the social awareness surrounding issues of violence against women. So the assaults – about principle, about feminist practice, about means, about goals – cut deeply across what once seemed a natural constituency of support. As one reporter opined, "it seems strange to me that women's groups, who would undoubtedly argue that violence is fostered by a society that condones it in thousands of small ways, don't seem to believe that the reverse must be true – that violence can be prevented by a society that condemns it in thousands of small ways" (Bula 1992).

As if there weren't enough concerns for the Monument Project Committee, there was also the issue of how, and whether, to invite men to work on this project. The committee had made a decision early on that if men wanted to participate and help out, fundraise and promote the project, they would be welcome. In fact, right at the beginning, Bill Tieleman, a journalist and political advisor to the New Democratic Party at the time, volunteered to help the committee write a direct mail fundraising letter, and indeed his efforts were used as the working draft for the first letter that went out. This fit with the initial decisions about how men could be involved, as advisors and as resource people. One committee member clarified the terms of involvement: "But, but [men] would never be invited to, to be an actual committee member. They wouldn't be a part of making any formal decisions about the monument."

Even with all the organizational incoherence and the setbacks in support, by the beginning of 1993, the project seemed more clearly on its way. Two December 6 anniversaries had come and gone. December 6 had been dedicated by Parliament as a national day of commemoration and education. Cate Jones, a woman with considerable fundraising experience, had been hired by the committee and had consolidated a campaign strategy for fundraising from foundations, unions, organizations and individuals. The committee was meeting regularly; city space had been (almost entirely) assured; the parents of the 14 women had agreed to support the project; and in spite of all the apparent rifts, still the project and other cross-country events had combined to keep the issue of violence against women in the public eye. The committee had begun to raise money in earnest, and it seemed that the end of the project was possible to imagine.

But there was still the problem of how to reconcile the project with its location. In a letter written on May 20, 1993, Chris McDowell wrote to the other members:

> I've been in the situation of "explaining" to women of the Downtown Eastside why we are naming the 14 women murdered in Montreal and none of the women murdered here. What I've said is the 14 women symbolize the importance of remembering the loss of individual women ... they say that that is not true, the 14 don't mean that to them and we should name Downtown Eastside women. The problems are obvious really ... the class and race differences get in the way of [the Montréal women] symbolizing poor women, women of colour, et cetera.
>
> Lately, I agree with them. When this project started we did not know that the 14 women would be remembered so well. We did not then know that so many people in Canada would respond so strongly to those murders.[9]

And then, two months later, another surprise: On July 19, 1993, the Park Board approved the monument in principle and officially named Thornton Park as its selected site. Design specifications for the monument were clarified, using much of the language that had, up until this point, been used in fundraising letters and letters soliciting statements of support to demonstrate that, indeed, there was a "community of interest" in the monument.

For some reason, however, the reaction to this news (though the details of the monument project had long been in circulation) galvanized vicious resistance. This resistance took several forms, and involved for the first time many new, articulate and socially prominent people who declared themselves publicly against the project. Most of them noted the inscription:

For all the women murdered by men
For women of all countries, all classes, all ages, all colours.
In memory and in grief
We, their sisters and brothers, remember,
and work for a better world.

9 Simon Fraser University Archives, Women's Monument Fonds, F-101-2-0-2, letter from Chris McDowell to Committee members, May 20, 1993.

The phrase "murdered by men" is, of course, explicit. True, the texts of most memorials don't cast blame, but to be unspecific about the male violence in violence against women was, in the eyes of the Monument Project Committee, to be unacceptably ambiguous. There had been no ambiguity for the 14 Montréal women – they'd been murdered by a man *because* they were women. Their murderer had even systematically separated the men from the women so as to ensure he killed only the women (although later he did also injure a number of men). He left a note explaining his reasoning, and listed several other women whom he believed deserved to die. However, just as in the original debate following the Massacre, the right to speak explicitly about the crime was, again and again, assaulted. A city councillor said he was "offended" by the wording; another called the inscription "severe and misleading." Apparently it was okay to say that women were being murdered, but not who was doing it.

The arguments against the monument presented in letters to the editor, in opinion pieces and in letters to the committee were circular, relentless and very, very familiar. The proposed wording of the monument was refuted as spurious: "There is no quantification of 'all the women murdered by men.' It is in keeping with the recent report from the Canadian Panel on Violence Against Women. This is the one with the poisonous, unqualified opinion that virtually every woman in Canada has been abused. Yeah, sure for $10 million, we get a 'survey' that conveniently reinforces every stereotype of male behaviour" (Rayner 1993). There was an attempt to deny overwhelming empirical evidence: "And in this era of gender equality and understanding, why are we taking a step backward and stereotyping all women murderers as men?" (Jang 1993). Some tried to change the subject: "The point, of course, is this: We have established as acceptable in this country that while all sorts of moral and legal barriers protect the 'disadvantaged' from criticism, slander and abuse of any kind, the supposed 'advantaged' (meaning white males who earn a living and aren't gay) may be maligned at will" (Byfield 1993). And there was confusion about the prevalence of violence in terms of gender: "Men are five times as likely to end up dead as a result of violence ... Amazing, isn't it, that people most often victimized by violence are men and the violence is most often inflicted by other men. If men kill men, maybe it doesn't have anything to do with gender ... Gender categories prove themselves as irrelevant as racial ones under even the slightest inspection" (Lett 1993).

But more worryingly, there was inflammatory ranting:

> The radical feminists have disgracefully diminished the slain women by expropriating their deaths for their own ideological purposes, while at the same time rehabilitating their killer and absolving him from his awful deed by insisting that after all he was just like other men. They happily let one man off the hook the better to hang the rest ... The man-haters know exactly what they are doing. They, and the larger movement of which they're a part, are not liberals. They intend to re-define common sense. They hate Western Christianity. They hate Western capitalism. They hate Western individualism. They think in groups. The state is their pal ... It's the wider public that doesn't know what it's doing. It's disarmed. It's confused ... It loses every case against the moves of the sexual Stalinists. (Lautens 1993)

There were dozens and dozens of such pieces in local newspapers over a four-week period. As a direct result of this kind of response in the press, phone calls began to flood the Park Board office, and as of July 26, telephone support was 30:1 *against* the monument project. Phone calls into the monument office – both for and against the monument project – had also increased in number. The committee was notified that several previous donors and supporters were "getting cold feet"; at the same time, however, unsolicited donations had increased – more money had come in since July 19 than in the rest of the year in total. But other events that month also affected public perceptions: one of the organizers was called at her home and personally threatened, and a bomb threat was made against the project office.

Such incidents drew the most attention, of course, but a selection of other voices indicates a shifting discourse around the whole topic of violence against women. Responding to Lautens' "feminist expropriation" column excerpted above, a reader mused:

> Mr. Lautens is vehemently opposed to the erection of a monument honoring women slain by men because he is concerned that it is sexist

> and "man-hating," although blatantly misogynist pornography and billboard advertising that uses sexualized images of women's bodies to sell everything[,] from gym memberships to stereos, freely circulate in this city.
>
> Is he afraid *the monument could help redefine common sense* and thereby provoke some men to become proactive against male violence against women? Certainly, by belittling attempts at making the public aware of violence against women, Mr. Lautens is saying in effect that he is not prepared to let women take control of their own voices. Just who is being sexist? (Duthie 1993; my emphasis)

The monument organizers also tried to enter the debate constructively. In an essay, two members of the Monument Project Committee reiterated the reasons for the monument and its origins, and wrote out the inscription in full. They suggested that some of the controversy was due to the perception that such explicitness over the violence ("by men") meant "we are not honoring dead women, but dishonoring living men." The essay quoted Park Board Commissioner Dermot Foley, who had asked openly in a meeting, "Who is the group this [inscription] is offending?" The essay detailed the reaction:

> The response of almost everyone in the boardroom – women and men, staff, media, project supporters and strangers – was spontaneous laughter. Foley went on to say he wasn't being facetious, but that it was obvious the only group being tarred was male murderers.
>
> That the inscription inspires unease in some people, women and men alike, is to be expected – senseless death is not an easy subject. Only by facing up to the twisted reality of what violence against women can lead to, will society be moved to change. (Phillips and Jones 1993)

Of course, there were also thoughtful men and women who spoke out. One man, who had come to the project as a volunteer, was reported by Cate Jones as saying, "If we want men to donate, we should make sure our language is inclusive." Jones continued, "As someone who witnessed his father abusing his mother he would really like to donate, but said he found the language on the dedication excluded him as a male." Another reflected, "'Women murdered by men.' Honestly, I really believe that made people so uncomfortable. That ... even if you're a feminist for ten years, you're not necessarily that kind of feminist that could say that ... Because that split the feminist community like mad. That wording." Another added, "But you know it is important to make a statement that lasts ... because people lie so much." The debate raged back and forth across newspapers, airwaves and dinner tables. Two women who had previously donated money wrote a thoughtful letter withdrawing support and requesting that their names not be inscribed as donors because they thought the inscription wording now represented a "wasted opportunity": "We don't object because [the phrase] isn't true or because it will offend men or because it's too radical ... [We object because] the inscription leaves out not only women who are not murdered, but were beaten, raped, threatened or abused as children, but it also leaves out women who are abused by other women ... At this monument, of all places, do we want women survivors of *any* abuse/violence to feel invisible?"[10]

There were suggestions to alter not only the inscription but even the form of the monument/memorial: why not make it, for instance, a rose garden? One letter writer responded: "The trouble with rose gardens is that their beauty lulls one into a state of peaceful acceptance and, in the bee-filled heat of a summer afternoon, a kind of forgetfulness. But how can one forget what our society has barely begun to acknowledge?" (Drake 1993).

Many suggested changing the inscription, making it refer to "all victims of violence," or "all victims of violent crime." It was in this atmosphere that the Monument Project Committee held a meeting mid-August that addressed the discussion around what to do regarding the inscription and the use of names, as well as the bomb threat. On these issues, the committee's minutes are reproduced here in full:

CATE: talked about feedback on dedication.

MARGOT: feels that listing all women adequate; the politics of identity is a shifting discourse; has talked to lots of people who want more women

10 Simon Fraser University Archives, Women's Monument Fonds, F-101-3-0-7, letter received from A. Vrlak and S. Hornstein, August 17, 1993.

named; likes idea of having names of women murdered between December 6, 1989 and when WM [women's monument] unveiled.

CHRIS: wants to focus on "by men"; it's the burning issue; Chris has not changed her feelings on it; she had been vacillating on it but she thinks it was because she was afraid; wants to be forced to change it; public debate has hit on why it's needed; stands by dedication as a whole.

ELINOR: feeling very tired of it; has thought about changing it, suggestions on strategy; check stats on how many articles published whole dedication, etc.; should use bomb threat to our advantage.

LORNA: keep it; can't back down now try to draw stronger connection between women murdered and 14 women; list 14 women, say these women were murdered December 6/89; since that time "X" number of women were murdered by men; this is taking back space for women and women's issues; look at how we win people back on a symbolic issue.

JANINE: more determined than ever to keep "by men"; concerned about letter from Joan; doesn't favour using bomb threat.

CAROL: doesn't want to use bomb threat either; supports "by men."

ANGELA: supports "by men."

CATE: supports "by men"; major political battle ahead; we have to put a plan in place.

MARGOT: "pissed off" at men writing letters that say we'll use vandalism, etc. as proof of our cause; not talking about bomb threat is another mechanism for silencing us; we should talk about it later in the project; it should be part of our history of the Monument now ... discussion around how much energy should go into this; can the park board take this away from us?

LORNA: asked Chris what if it's "by men" or nothing?; can we go with an inscription at all and in the artists' guidelines talk about the inscription we wanted, say it was turned down and see what the artist comes up with?

CHRIS: wants inscription to stay because it's the voice of the women who created it; she wants "by men" but can't imagine getting it.

ELINOR: it's important that we get a lawyer on board in case they pull the rug out.

AGREED To keep "by men" and take it to the Park Board
Do we go with an alternative?
Do we leave the inscription up to the artist?

LORNA: don't underestimate the ingenuity of people submitting; connect the dedication with the 14 women.

JANINE: don't react to the public – let them battle it out; keep on with our work; re-establish relationship with Park Board Commissioners; we've defended ourselves enough in the paper; we have to get away from the defensive position.

MARGOT: we'd betray ourselves if we changed the dedication; we're making history here; if Park Board bends *that* needs to be the story; need to phone people to phone in support; what we're asking for isn't outrageous – when we get turned down, we expose it.

ANGELA: can we put the WM at Cap College with our dedication? Can we get war veterans to stand up for us? If we lose "by men" at least we'll still have a monument.

AGREED If we can't get "by men" we ask them (Park Board) what they'll accept; get their list and then say we'll take their suggestions back to the WMP [women's monument project] committee.[11]

In the end, the committee did not back down. After a summer of relentless attacks and high tension, September brought with it a slight shift. One editorialist travelled north to a small town in BC to assess what violence against women looked like outside an urban context, and wrote back:

> In Vancouver, nice guys fret over whether words on a plinth might damage their reputations. Around them women are beaten to death, raped, assaulted, verbally abused and psychologically terrorized by the tens of thousands.
>
> One in ten women gets this treatment, says the most comprehensive survey ever taken. You figure it out. About 750,000 women in Greater Vancouver points to 75,000 casualties.

11 Simon Fraser University Archives, Women's Monument Fonds, F-101-2-0-3, minutes of the meeting of Women's Monument Project, August 14, 1993.

> ... Women are targets of directed violence. It's their femaleness that's assaulted. And we all know this violence is not about sex, it is about establishing who has power. It is intended to invade and occupy. (Hume 1993)

An article in the feminist magazine *Kinesis* noted that even WAVAW had recanted its earlier objections: "the inscription now clearly recognizes all women murdered by men." The Vancouver Status of Women acknowledged that the monument "is not what we identify as a priority ourselves, right now," particularly in light of the "tremendous racist backlash right now against women of colour and Aboriginal women," but added that every contribution women see fit to make to end violence against women is welcome: "The monument project aims to show people that these things occur. It's going to be visible and that is good. It will be a reminder" (Bouchoutrouch 1993).

The debate was not yet over (arguably it is still not over). Editorials, cartoons and arguments appeared regularly in the media, especially over the next year. Privately, one of the Park Board commissioners and the committee met to resolve the inscription issue. They emerged from that meeting having preserved the inscription in its entirety, but inverting two lines: the inscription now began with "In memory and in grief." Park Board Commissioner Nancy Chiavario noted she personally thought the resolution had been possible because "the project was a right project. Otherwise we wouldn't have seen the answer." By December 1, 1993, the Park Board had unanimously approved the application, and the design process was officially launched. In one way, the Monument Project Committee's work had been simplified; it other ways it was just beginning. To run the national competition, the committee was hoping for guidance from the Public Art Committee of the Park Board, but they were disappointed. The Park Board's assumption, that a national grass-roots feminist design competition could be completed in three months was, in the opinion of one committee member reflecting on that period, "honestly, so wrong of them. So profoundly wrong." In the end, the competition took four months (though it could just as easily have been six) in order to make the process as fair as possible.

Part of the difficulty in organizing the competition had to do with the intended openness of the process. First of all, more than 4,000 calls for entry in English and French were sent out across Canada to the media, educational centres, women's centres, First Nations' media and to magazines and newsletters about architecture, landscape architecture, engineering and the visual arts. But as one committee organizer asked, "How long does it take you to get the call out? How long before it gets picked up by the media? How long before it gets to your town?" It is important to remember that this competition was held in the days before widespread web availability: the calls for entry were sent out by post. The committee also intended that there be a stage in the entry process "where artists, people who wanted to submit ... could write down their questions, submit it to the jury, we had to answer those questions, and get it back to all the people who had inquired, all at the same time, so that everybody had the same information." This created a kind of logistical nightmare, as Chris clarified: "Your call goes out, and [artists] submit and get a package. So then, everybody who submitted to get a package gets this letter ... with the question and answer. Then there's more questions, then there's more question and answer ... [yet] the public art committee was adamant ... about this three month thing." Once again, the committee proceeded despite less than ideal circumstances.

The design call went out with an explicit rationale for why only female artists were asked to submit proposals. Ninety-eight entries were received. The competition was to be juried by seven women from across Canada, all of whom had made significant contributions in the visual arts, literature, architecture or human rights: Nicole Brossard, Rosemary Brown, Wilma Needham, Maura Gatensby, Doreen Jensen, Haruko Okano and Irene Whittome. This was the first time in Canada that a jury had been made up of only women.

There were two stages to the competition: The first stage was a blind review in which only the design was judged and the proposals remained anonymous; three finalists were chosen and given development money to further their proposed projects. During this second stage of the competition, the jury met with the finalists and discussed their entries. The winner of the competition was chosen following these meetings. Yet in addition to the official competition, the submissions were featured in other fundraising, education and advocacy events. Maquettes of the three finalists' designs, as well as 16 honourable mentions, were involved in four showings entitled "The Women's Monument Exhibit" (in Toronto, Calgary, Vancouver and Surrey). There is a strange "What if?" tangent to the selection of the final design. Artist Susan Point was originally one of the three finalists. Now a highly regarded Coast Salish artist, Point had withdrawn her participation in the competition due to a death in her family, fearing she would not be able to work within the specified time frame. If the winner had been a First Nations artist, would this fact have altered the debate around placement? Around the right to erect such a monument? Around the right to criticize it? Would a winning entry by a First Nations artist have affected the time it took to fundraise or the public's reception of the project?

In October 1994, the Monument Project Committee announced that the winning design was Beth Alber's *Marker of Change*. Alber offered an elemental design – a wide circle of 14 benches surrounded by tiles naming the donors to the project. With the project's details moving anew into public discussion, the press once again featured many shrill and inflammatory reactions. Though the design was now clear, the pressures of organizing the competition under these circumstances, as well as the enormous time commitment (meetings often lasted several hours and, because all committee members worked, had to be held on weekends or evenings) wore on the committee. One member who lived through the whole period reflected: "I think that the group that had so much conflict – and it's painful even to think about that group – was a very courageous group, and took on all those arguments over and over ... because that was during the absolute shit coming at us about everything. You know? Everything. The wording ... of the dedication, the women-only competition ... you know. Even being a women's group is suspect, right? And that was a great group for ... arguing that publicly and insisting on certain points of view."

In time, new voices of support rose up. One of the most moving of these was that of a man who worked with the group and whose opinion piece appeared first in the local *Vancouver Sun* newspaper and was then reprinted nationally in the *Globe and Mail:*

> To deny the phrase "murdered by men" is to ignore who is raping, who is killing ... What are the consequences if men insist upon removing the "male" from "male violence"? In effect, it is to suggest that violence is some natural phenomena [sic], like the weather, which Canadian men and women have to endure ... It is men ... And they are committing these atrocities under cover of the reputation and good name of men who are decent, non-violent and law-abiding ... I realize now the simple and painful truth of that inscription. *It is not about collective guilt, but collective responsibility.* (Gale 1994; my emphasis)

Slowly, the public represented by the monument, and the movement against men's violence against women, was redefined. The focus was not anti-male. The public the monument claimed to address was full of men, women and children. The money for *Marker of Change* trickled in again, but now the really serious work of fundraising began, and that meant further organizational changes. The committee decided to become a closed collective. As one advocate recalled:

> We had to stop sitting around the table and, and just dealing with people's different ideologies and philosophies on certain things, because we had to get out there and make the money ... Like VanCity [a local credit union] said, certain criteria had to take place before we could get their money. And that was a pretty substantial chunk. And once we had that, then we could go to other people and say, "Look, VanCity's given." So there were all these things that had to come into place. I think that was part of the motivation for saying that we thought, well, if women come on a volunteer basis to three meetings. Maybe that never became totally formalized, but we had talked about that. That if they showed ... that it wasn't until after a certain period of time, or maybe it was three months of being involved, that then we would invite them to actually be key decision makers.

All agreed that the pressures of fundraising changed the dynamics:

> I remember that we were drawing a big debt with Cap College. And honestly I thought so many times, I just felt like smashing my

> head on the wall. That while we're drawing all this money, we're arguing for 10 months about the wording of the first direct mail letter. It was just unbelievable. And I just couldn't believe that it was happening. But certain people in the group ... whoever they were ... couldn't allow things to resolve ... they just couldn't. And I think there was a resistance to fundraising. That they wanted to be there for the ideas part, but for the serious fundraising, when the serious fundraising started, they left. In hindsight, it seems some of them may have been there more to influence the competition than work with the fundraising. You know, in retrospect.

When fundraising had to get serious, the committee came up with the idea of a "film-a-thon." At this point, the committee had only three people. The committee set about recruiting new members (including women who then stayed with the project all the way to its installation), and two men. Krista, one of the women who stayed on, said:

> When I joined, I didn't meet the committee. I met with the people ... who were going to be doing the film-a-thon. And for months, or, actually, I don't know how long, we had this ... very committed group of people, and we couldn't do anything. I hated the monument. Or the Monument Committee, I should say. They were like the Wizard of Oz! It was, like, everything we decided, we'd have these incredibly intense meetings, we'd have all these great ideas ... we have to run it by the committee ... Finally we were just so fed up, we were just like, "Look. We need some independence here if we're running this thing!"

Over the next few years, the efforts and setbacks around fundraising were a source of constant tension, but the minutes of committee meetings also show that extraordinarily positive, unexpected events occurred as well. A Québec quarry firm contacted the committee to say it would be honoured to donate the granite slabs; Sumas Clay likewise donated the tiles that would encircle the monument and then later helped a group of volunteers stamp them. The Women's Monument Exhibit went on tour, and the committee and its advocates followed a constant schedule of fundraising and press events. Suzanne Laplante-Edward (mother of the

"This is ... the most heartwrenching piece of writing ... because it is impossible to put her brief but meaningful life into a few short sentences."

— Comment from one of the parents of the murdered women[12]

murdered Anne-Marie Edward) encouraged and worked with each family to create wording for a personal tile for their daughter.

In mid-summer 1996, the committee held an "envisioning ritual" in Thornton to help raise money and to imagine the monument in place. Even though several members told me they felt like "idiots," a group of about 10 took string and stakes with them and actually created a little monument one night in the park. This event had at least two other unintended consequences. First, the police came asking questions; as it turned out, someone had robbed a gas station nearby and the police had assumed the thief had run across the park. The women laughed as they told me the anecdote, noting, "So the money was coming through the park, it just wasn't coming to us!" There was no laughing about the other event, however. At a "very intense" moment in the ritual, several men approached the group. Though they seemed simply curious at first, they became incensed when told that this was a private ritual, and to please let the women be. Three of the men became very hostile, tore down the little monument and threatened the women. The women left the park unharmed, but set a precedent that evening to always leave Thornton Park with a particular set of ritual gestures and gifts, a practice they still continue.

Finally, in 1997, as it appeared the monument would actually be built soon, the committee undertook the task of determining which languages would be represented on the benches. Initially, the committee decided that it would choose seven major world languages and leave seven benches without an inscription. In keeping with their commitment to

12 Simon Fraser University Archives, Women's Monument Fonds, "Granite and Circle of Donors/Tiles." Part of a package referred to as "Sept. 30, 1997 Summary of Project Major Events, People, Aspects and Issues, handed out to media."

open and consultative practice, the committee developed a questionnaire and circulated it widely across Canada to determine both a process and a decision-making group. This group, made up of the design jury, the project committee, Carol McCandless (from Capilano College), Suzanne Laplante-Edward (mother of Anne-Marie), and Beth Alber (artist), decided on five languages by consensus, but the group's right to choose was challenged by a woman at Douglas College. She wrote back, "You have no right to pick all the languages." The committee members' reaction surprised even themselves: "Oh, we were so happy. It was like 'She's right! We have no right.' We're going to ask the Black Women's Congress to pick the African language. And then we, we asked these Native women to pick the Native language. And it was like 'Phew! Thank God!'"

According to several committee members, it was within the context of languages that the first "real" conversations with activists in the Downtown Eastside took place. Up to that point, some of the committee members said "it felt like a really trendy thing" to be *against* the monument, as though it were somehow being imposed on the area. The irony was that through the long process of public education associated with *Marker of Change,* "the issues of the Downtown Eastside became more revealed, and there was more media and public attention." The question of languages provided a direct, pragmatic vehicle for discussions that were more difficult. Lianne and Janine described this process:

> I remember there was some dialogue that started to happen around the languages. "Well, just what is it that you want us to do? And why do you want us to do this?"
>
> There's a whole thing around this, though, that was unsolvable. And unresolvable, that you just have to [pause] just admit ... [We] asked them, if there was a viable project happening, could we do a last direct mail, after we're finished fundraising ... but there was no solid project emerging from the Downtown Eastside. But ... it's just so painful. Because, ultimately, what we were faced with, was women saying, "You know, my daughter, or my friend is murdered here, and somebody else's name is going to be on your benches, who lives 3,000 miles away ... why aren't you naming local women

> on your monument?" And ... there's actually an answer to that question. But, but at the same time, there is no answer [pause] except that, yes, there should be another one, and yes, it should come from the Downtown Eastside, and it should address women lost in the Downtown Eastside. And ... we couldn't, we couldn't transform our project into that.

Yet the placement of *Marker of Change* served the idea of such a monument-to-come. Lianne mused that if *Marker of Change* was at "Oak and Sixteenth or something ... it wouldn't bring that kind of attention ... But [*Marker*] is ... sort of a bridge, in a way. It's a way of leading to their project." But the bridge was still very uncertain. An anonymous insert was placed in the Carnegie Community Centre *Newsletter* of June 15, 1997, including excerpts from a poem identified as being written by Barbara Gray, with the note: "If it is accepted by the community in general, Barbara will be asked for permission to use [the poem] as the dedication" on a tile specifically to honour local women. It read:

> To all women on
> The downtown eastside:
> We come together
> To begin the healing.
> We have trudged
> These paths before
> Too many times.
> We have had enough –
> Enough violence
> Enough beatings
> Enough stabbings.
> We shout we scream
> No more.

In the next issue of the publication, Barbara Gray herself wrote, "I do not support the December 6th Women's memorial, their committee or anything and anyone connected with this project." In the end, the committee acknowledged it was unable to successfully collaborate with the Carnegie Centre (the city-funded community centre located at the neighbourhood's main intersection of Main Street and Hastings Street) and the Downtown Eastside Women's Centre to create a tile in memory of the local women killed. Instead, with the assistance of two women who are poets and supporters of the project, the committee approved the following:

> IN LOVING MEMORY
> OF THE WOMEN KILLED
> ON VANCOUVER'S
> DOWNTOWN EASTSIDE
> SO MANY WOMEN LOST TO US
> WE DREAM
> A DIFFERENT WORLD
> WHEN THE WAR ON WOMEN
> IS OVER

In July 1997, the *Marker of Change* organizing committee broke ground and began the process of installation. Vince Stolon, an elder of the Musqueam Nation, was invited to Thornton Park to offer a blessing for the groundbreaking ceremony. He found that one member of the committee was crying profusely: a friend of hers had just been found murdered. Stolon would not permit her to stand within the circle while the ground was dug – a decision that bewildered the woman and infuriated others, who saw it as a denial of the very point of the monument. It was a painful moment. Some imagined that, given that the spirit of beginning something has consequences, perhaps Stolon was not acting in denial but for the protection of all concerned – the earth, the dead woman, the people present and the "spirit world."

The installation period, like so many other moments in this monument's development, was fraught with unforeseen obstacles. A city strike meant that all building permits and much city equipment were frozen for several weeks. Asked how this affected her, *Marker of Change* artist Beth Alber, who was visiting from Toronto specifically to oversee the installation, sighed, smiled, and said, "Something in me knows this monument wants to be in the world. It just does. I trust that." And sure enough, the strike was finally settled just in time for the complete installation to begin before Alber had to return to Toronto. The monument's design features 14 benches arranged in a perforated 300-foot circle. The design is visually reminiscent of ceremonial sacred geographies, from standing stones to healing circles. It is a non-hierarchical form representing a continuum and suggesting a protected or consecrated space. There are a number of other subtle design elements. The solid mass of the stone benches and each bench's length of five and one-half feet suggest a collection of coffins. Alber, an artist highly attuned to the qualities of the materials she uses, originally wished to use construction-grade granite to underscore the fact that the murdered women were engineers, but when tombstone-quality granite from Québec was donated, Alber found this gesture so powerful that she adjusted the design. The gloss and the hue of the donated granite are thus both suggestive of cemetery stones. But Alber's choice of granite itself tells a story: Alber wanted

Marker of Change | The ring is made up of tiles on which are written over 6,000 names. Notice also the sandblasted "gash" on the bench top, which artist Beth Alber indicates "is a place for water to gather, a sign of life," though she notes that "the scars will also never heal." A scar is a part of each bench. | *Photo: Greg Ehlers*

Marker of Change | The benches are intended to be deliberately suggestive of coffins and are the approximate size of an average North American female body. | *Photo: Greg Ehlers*

materials that were taken violently from the earth, wishing to learn to work with the energy and loss that came from the stone. Indeed, she wanted to emphasize the loss, rather than submit to the sombre acceptance of these deaths. To do so, she "gashed" each bench with a long, sandblasted form, unique to each. She suggests that this makes a place for rain water to gather, as water both renews life and collects reflections. Scars, however, will never disappear.

The monument is intriguing in terms of its ability to communicate with viewers. It is open and accessible; there is no point of domination, and it can be seen clearly from a long distance, though it blends in completely with the other benches of Thornton Park. Each bench on the inside of the circle bears the name of one of the women killed in Montréal in 1989, as though these women, who were in many cases looking at each other as they died, are now memorialized in a similar configuration. On the outer circle, the inscription is written on seven of the fourteen benches, in one of seven languages. Around the benches, as noted in the fundraising brochure, "a continuous ring of paving bricks behind the granite forms are positioned in the ground with the names of the contributors handstamped into clay. The circle of donors acts as a frame – a protection, a caring gesture. An orientation stand is positioned outside the circle with Braille and raised lettering for people with sight-impairment."[13] Beth Alber originally wanted only names on the monument – she felt silence spoke most eloquently. But in time she agreed to the inscription, for what are perhaps paradoxical reasons: "[Every time I hear the inscription] it still makes me bristle; I think that's exactly what it's supposed to do."

In December 1997, eight years to the day after the Massacre, the dedication process began with a group of local First Nation women marching through the Downtown Eastside, drumming, singing, and praying, all the way from Carnegie Community Centre to Thornton Park, where a huge crowd waited. The day's speeches proceeded, braiding anger with grief, and intimacy with formality, in a variety of languages, cultures,

13 Simon Fraser University Archives, Women's Monument Fonds, F-101-5-0-10, "Why Thornton Park?" Part of a package referred to as "Sept. 30, 1997 Summary of Project Major Events, People, Aspects and Issues, handed out to media."

gestures and silences. The diversity, enormity and dignity of the crowd astonished even the organizers and all of the family members who came to see the monument for the first time that morning. My own account of the day recalls some details:

> The whole event was taking place amid the miasma of the death of Reena Virk, at the hands of seven girls, in Saanich, BC.[14] Suddenly, it seemed, there was a new and terrifying shape of violence against women ... It was an exceptionally cold day ... [and] I feared lunatics and violence. I expected about 35 people ... As I came off the Skytrain, I heard drumming, and saw the crowd in the thousands, and I felt tears of excitement, energy, gratitude in my throat ... I saw hundreds of men, all races of people, dogs, infants, elders. There were several speeches ... at first, there was no mention of male violence, only of loss of these young women – so many young women – lost to us ... then Suzanne [Laplante-Edward] spoke, beginning with how she speaks often to "the girls," who give her energy, and crediting them with the beautiful bright day, so unusual for December in Vancouver, saying it was their brightness brought near. She spoke quite personally, and quoted from a long piece written by a man, in Québec, a few days after the massacre ... The line I still remember is "How is a man's fist in a women's face only a woman's problem?" ... [Another speaker asked us] to form a circle around the benches.
>
> There were so many of us, it became two or three or maybe four rings. We faced each other over the sarcophagi, the slight rise of the earth green and frosty under our feet ... I found myself holding hands with strangers; at different points each of us cried. We stamped our feet against the cold, looked at each other occasionally, but did not speak. Music, written especially for the day, swelled over the crowd and we fell silent as the unveiling started across the circle from me. I knew from Beth it would be the bench for the first woman killed, as Beth had ordered and worked with

14 Subsequently, both male and female participants have been tried and convicted of the murder.

Marker of Change unveiling ceremony, December 6, 1996 | This photo was taken before the drapery was taken off each of the benches. Notice that the distance between the benches is too great for a conversation to take place; the viewers must physically enter the monument's spaces in order to converse with each other. | *Photo: Tim Matheson*

the benches so as to honour the times and spaces of the moments these women had died. Two people lifted the shroud over each bench, folded it slowly, and laid it with the others on outstretched arms. A lantern was carried around the circle, lighting one at each bench. To complete the circle with such reverence took a long time – far longer than it took to kill these women. It was an extraordinarily long time for a crowd to be silent, a long time for strangers to touch. At the end, we were invited to place any things we had brought in the part of the monument most meaningful to us. In a matter of moments, each bench was covered with flowers, candles, messages, as were many small places in the grass. The circles broke, and people re-massed in to the heart of the monument, some going to the south side of the park for the tent with food, warm drinks, seats, a petition in favour of gun control.

> There were more speeches. The mother of Cheryl Ann Joe – whose death is now commemorated as part of the Feb. 14th march. The woman who was chief engineer on the monument, talking about the night of the Massacre, and how she phoned the only other woman engineer she knew in BC, and how they cried ... I watched as the engineer faltered, her voice breaking when she spoke, about giving birth to her own daughter and how that crystallized her hopes *and* her fears – Lianne came behind her, put her hand mid-back and rubbed up and down her spine – energy, warmth, the message she was not alone, was only the voice for many. That gesture seemed the whole essence of the monument to me.
>
> ... I happened to turn the moment Chris and Suzanne passed Suzanne's daughter's bench. They saw each other as I saw them, and collapsed into a long embrace behind the mound of flowers. It struck me then – as extraordinary as this monument is, as its development has been, as brilliantly as the day had gone, as vibrant as the atmosphere felt – strong, clear, determined to do better in the world – soon twilight would fall, the benches would stand in the night alone in the cold and darkest hours. I watched these two women, sobbing and smiling and holding each other, who had come to know each other for all the wrong reasons and struggled so hard across years and a continent to make this moment. And at the centre of it, Suzanne's "girls" are dead.

Since Installation ...

Since the unveiling, of course, violence against women has not stopped, but the committee's aspirations that *Marker of Change* would be an interactive monument, a place for reflecting and commenting on such violence, have been realized. There are both formal and ad hoc uses of the monument. On the key ritual day of December 6 there are a variety of activities. The committee members usually clean the monument the weekend before, carefully tending the site and washing each bench. On December 6, many of them go first thing in the morning, and place candles and flowers and food on each bench, walk the circle, spend time there in silence or in offering a meditation. Over years of observations,

I have noticed that a pattern has developed over the mornings and early afternoons of December 6. Many people come and spend a few moments there. Their styles differ, as shown by my field notes: "A tall man, white-haired in dark, sharp clothes with an umbrella, walking by himself on the donor tiles, counter-clockwise. And pairs of women and men, hovering, pausing, reading a bit. Some two dozen over the morning." By afternoon, there are usually more:

1:45 PM media person, plus two, then another – women – lighting candles, some interviews. People tend to stroll around the outside rather than walk across ... This reticence, this gentle negotiation is different than I expected. There is something mediating about the site. It is liminal. One is not sure of preferences. Do people want to talk, or not? Are they praying? Crying? During all the filming, four of us just watch. Another goes by checking trash cans.

1:50 she is back, walking slowly with her bike. She is relighting the candles. Small acts of attention. Young boy running around on the donor tiles, like a game. Two more women come, with flowers. Appear to make a small comment/ritual before each bench.

2:10 another white-haired, blazered man at the plinth ... he stands like he's at an altar, his arms resting open on it.

2:21 blind man with a cane and a woman walk and light candles at three benches, then leave.

Sometimes I am amazed at the conversations people strike up with me, particularly men. My notes reflect my own wariness and ambivalence:

> As men have approached at the gatherings, marches, vigils, observations, I feel terribly caught between wanting to welcome them – feeling their presence and attention is absolutely essential – and being frightened that, perhaps, this is one man who will 'go off.' It happens. I've known enough vets, enough psychotics, enough good people that became ill, and isolated, and "lost the line." Any attention seems to draw more attention ... yet the presence of the

> circle seems protective. It occurs to me when I talk to men to be concerned – what if this sends them into a rage? What if they are men who, though abusive, are also charming? One man first sees me sitting on a nearby bench. He walks over to the plinth, looks at the inscription, walks closer by a few of the benches, then comes to sit beside me. He introduces himself (Larry), and asks, "So what do you make of this?" I say something vague and non-opinionated, and return the question. He looks at me for a while before speaking. "Until people come to a certain level of consciousness, they have no choices. After that, they do. But the guy who did this ... [he waves at the monument] ... didn't." What is Larry saying? The guy who did this had no choice but to kill? But Larry seems calm and we talk a bit further. Suddenly he takes my hand and places it on a (healed) wound near his knee "to show you how the pin has slipped" from some botched surgery. I am struck again that at this place, strangers touch. But this encounter paled alongside the quiet alarm I felt later that day as a man easily twice my size came up close behind where I was standing and suddenly barked, "Sad reality is – if it were eleven guys who'd gotten drilled ... this would've happened a long time ago." The miscalculation (why eleven instead of fourteen?), the insensitivity of being a big man suddenly far too close to a small woman, the loudness of his voice, the verb, all jarred me deeply. And the ambiguity of what he was saying – what would have happened? A monument? Men are killed all the time, and most don't get monuments. What was he saying?

Also on these days, I have witnessed that two teachers from the nearby ESL course centre bring their classes "every year. It is so important. [The students] are always shocked, really shocked. People can't get guns in their countries (Japan, Korea), so we talk about what violence against women is like – other kinds. After, the women feel a bit more afraid, so we talk about that."

Occasionally, quite extraordinary encounters happen. On December 6, 2001, in the late afternoon, I met a woman at the plinth – she is a Métis who works with the UBC Aboriginal Law Project, and who had driven from Saskatoon a few months earlier. The previous night, she had seen a documentary about the monument, and had felt "called" to come. She pulled from under her coat a small, heavy iron pot – her mother's smudging pot, given her for this time away from home – and some sage she and her partner had picked along the drive. My notes recall what happened:

> We talk, she asks me to smudge with her. Muddy ground. She touches me. We smudge ourselves, then light the candles and smudge each name as we squat by each bench "Prayers for ..." She doesn't want to go backwards in the circle, and thanks me when I do to fetch candles. She says, often, when things are not just right, 'must be meant to be.'
>
> She tells me she is 'sick of white rich men' doing violence to native women in Saskatoon. I tell her it is very similar here. We kneel, light, say the names together, smudge.
>
> At the last bench, Randy appears. Shy, well-muscled, not apparently high at the minute anyway, and he begins tales. Says he's been trying to keep the site nice. (In fact, there are flowers, candles, oranges at each bench and around the circle, with some in the middle). Says he's going back to Red Deer. He says things like "ma'am." Mentions he's a carpenter. Says he doesn't want a handout. But he is talking, interrupting.
>
> I find myself irritated by how I cannot ever come to the monument and simply be still without getting hit on. However, we are out of matches for the last candle, and Randy ... figures out how to light it. Anna and I are both, I can tell, not sure what to do with Randy. I say I want to finish our ritual – Anna invites him to circle with us. Amazing – it calms his story, and he joins us, walking.
>
> As we go 'round they ask dozens of questions and I find myself telling them stories – amazed I know them, amazed I have joined the story of the monument myself. We talk about languages, silences. We complete the circle and it starts to rain. Randy tells us he is here because his tools are stuck in a locker in the bus station that he thought was one price, but it turned out to be a per day price. I decide I believe him, and decide to give him some money. I also tell him that it is likely when the events start at twilight, there will

> be food. When we go towards the side of the park, we see a double rainbow, and it makes Anna cry out with joy – it means much in her family. Randy spontaneously hugs us both, then leaves to check about the food. Anna hugs me and laughs. We say goodbye, and I come home. The next day, she sends me an email, though I did not give her my address, and she says she is glad I am doing this research, because I am a person who listens. I read, astonished, and take it as both instruction and encouragement.

Sometimes, the monument is used ritually on other days. The first public gathering called there, curiously, was for neither a Canadian nor a woman. It was a vigil for Bernard Slepian, an American doctor who worked with women on all kinds of health care and pediatric issues. According to the one who shot him, he was murdered because he also allegedly performed abortions. Other spontaneous responses at the site appear as reactions to the news about horrific murders or beatings, or to the events of 9/11 (a giant peace symbol was fashioned from flowers within the circle) or in honour of the Valentine's Day marches. (These marches, held every February 14th since the early 1980s, seek to commemorate and remember the missing and murdered women of Vancouver's Downtown Eastside. Though every year the route alters slightly, certain customs have remained constant: leadership for the march comes from female Aboriginal elders, who offer prayers and rituals at key sites, including the last places some of the women were seen, or the locations where their bodies were found.) Often certain benches are decorated with flowers, messages or candles. There are happier uses too. Carvers come to work there (as indicated by the shavings left behind). Once, someone unrolled a sleeping bag in the centre of the circle and slept long into the morning on a cold autumn day. In 2001, there was an Earth Day celebration called there (though the memorial was referred to as the "Women's Monument"), and in March 2002, for the first time, the official International Women's Day rally was called at the site. More casually, though people most often sit on nearby benches rather than on the ones in the monument, on occasion people will picnic or nap on the monument benches. Birds drink from the water that collects in the sandblasted scars.

The monument is placed directly across from the train and bus station. | *Photo: Greg Ehlers*

Once, there was (indecipherable) graffiti left on the plinth, but other than that, I have never seen the monument defaced.

Given its heft, its substantial and powerful physical presence, it is interesting how *Marker of Change* functions in terms of media. The organizers realized early on that the whole idea of a monument as a response to male violence against women constituted a historical first. They began taking photographs to document the process. The fact that the project was immediately given office space by Capilano College meant that files were kept from the very beginning of the project – press releases, fund-raising letters, financial records, phone logs, minutes and original correspondence, as well as photographs and video. (The archive, officially opened in Summer 2000, includes over 12 linear feet made up of nine boxes of documents and items). In 1994, the May Street Group (a filmmaking collective) approached the Monument Project Committee and asked to make a documentary; it began filming in earnest around the making of the tiles, continuing episodically all the way through the unveiling ceremony. At the time, no one had any idea how the Montréal events would affect Canadians across the country or how the events would be memorialized. As they've unfolded over the years there have been literally thousands and thousands of texts, films, images and art projects devoted to commemoration, education and action. Yet even among these, *Marker of Change* assumed an iconic status because it was the first intentionally national grass-roots response to the Massacre following the funerals. The national juried art process, the tiny fundraising events in small towns and large cities, the radio shows, the direct-mail campaigns and the public speeches all contributed to the huge broadcast "reach" of the monument as idea and, after selection of the design, as image.

Because the monument is so large and so understated as a design, the media images used to depict it are largely stylistic – they capture the essence and intention of the forms, but not a complete visual rendering of the whole (except in one aerial shot). Instead, images focus on one or a handful of benches, offer a detail of a name, a donor tile or an inscription or show one of the scars full of water. Rarely do these images show people, except perhaps as non-identifiable individuals whose comparative size suggests the scale of the art. The only element of the monument that is described or depicted in its entirety is the text. This is interesting, given that when people come to the site, they seem most often to walk the complete circle of it.

When pieces of the marble used for the benches were passed around as part of information and fundraising events, people seemed notably impressed by the colour, temperature and weight of the stone. But more often, news of the monument has circulated via text (as in fundraising letters) or other media. The documentary *Marker of Change* debuted in December 1998, and copies are now housed in hundreds of college and university film libraries, primarily throughout North America. The film is broadcast nationally every December 6 on television as part of the annual commemoration events. It has likely been viewed by millions of people.

Yet the monument's advocates remain curiously wary about the role of the media. As Chris McDowell put it, and others agree, organizing events around violence against women is tricky:

> It can go sour. And it can go to a point where people are just completely fed up. That's what we found out. Because we suffered through event after event, and fundraising, and, you know, the years of experience ... But it is so horrible to do events that have two audiences: the press and the people attending. And, who are the people attending? Are they number one? They're number two. And it's so painful to be number two. "I'm here as a body for the press. To see that a body came." And that's so awful. And that's what we discovered. That we did that kind of event, we lived to regret it. You know, we got used by other people, we used other people. Finally, you know, we tried to cater to the media, the media just shot us down. When it comes to violence against women, why focus on the media? They are part of perpetuating violence against women. You know, if only people would just do events that were about healing and social change for the people that come to them[,] and screw the press.

This sentiment was particularly strong when the committee talked about how it had prepared for the day of the unveiling. Claiming they had learned so much from the power of ritual that characterized the Valentine's Day march and various meetings they had attended with First Nations people, the committee members decided to have "[a] press conference because we didn't want politicians to speak at the unveiling ... so we created two separate events." On the day itself, early in the ceremony, a cameraman slipped from the rings of people surrounding the monument and walked into the middle of the circle to "get the shot" of those who were lifting a veil from one of the benches. Immediately, Chris strode across the circle and physically dragged him to the margins. Such clarity about the role of media versus the role of attendees was also obvious in discussions about the "hand-off" event (in December 2000), when, for the first time, the committee was not in charge of the official December 6 gathering. Instead, the ceremony was to be run by none other than

WAVAW, the very group that had nearly destabilized the whole project back in its early days. Several committee members commented that it was a great relief to "see someone else take that [the official event] over," because it "was very freeing ... you know, we've already ... talked about what we want to do next year. And it's personal, like, we don't have to worry ... we don't even have to publicize it." For the 2000 event, WAVAW had originally asked committee members to be involved. The ensuing exchange was not without irony, as related by a committee member: "I phoned around, and people were open, and I phoned [the WAVAW organizer] back, and said when we could meet: Monday, Wednesday night or Friday night. And she said, 'At night? You can only meet at night?' And I said, 'Yeah, that's right.' 'Oh, well that's no good' ... and she said, 'Can't you meet during the day?' I said 'We all work.' You know. Like, we all work. And, [the WAVAW person said] 'Well, we'll tell you what to do.'"

The entire committee was incredulous. Of all the community groups, WAVAW ("after just about publicly crippling us") had little claim, in the advocates' opinion, to tell them what to do. In the end, the committee members attended the events, as did I, but did not take part officially. As it turned out, the WAVAW-led ceremony included a whole series of actions staged for the camera, including Japanese drumming, speeches about welfare and a shouted, angry address about why events like this didn't draw as many people as football games. The WAVAW organizers chose to place themselves in the centre of the circle (rather than to one side, as the committee had always done). Camera spotlights were set up to focus on those in the centre, and attendees were left to choose between not circling the monument or circling it, but ending in some cases in various locations being either blinded by lights or hidden behind the speakers. The spotlights required huge, noisy generators that made it impossible to hear. The event, in short, was designed at least as much, if not more, for the media as for other participants. As one woman said to me, "You don't want to be, but you know, you hear women talk about 'I went to this thing and I never want to go to a feminist thing again in my life.' And they mean things like this."

The framing that results from such media presence at these ceremonies and from using these monuments in this way is itself quite alienating: it often highlights sensational violence while simultaneously failing to provide any context in which to understand or address it. For example, one December 7, I read a daily Vancouver paper cover to cover and counted seven articles concerning horrific attacks on women, murders of women and threats to women. None of them connected to male violence as an issue, though all these attacks were instigated by men. None of them connected to the memorial services the day before across the country, even though an article about those services appeared in the midst of the others.

The CRAB Park Boulder

To understand the second monument – the CRAB Park boulder – requires another introduction first, to CRAB Park itself. The boulder was dedicated 10 years to the day after the park was dedicated; it is inconceivable not to link the two events. As noted, there is a density of occupation and a scarcity of shared public spaces in the Downtown Eastside. CRAB Park represents an extraordinary victory of residents in even imagining, let alone securing from developers, acres of public land that give local residents access to waterfront. Thus the park and all its features constitute a kind of running conversation about property rights in the neighbourhood. There have been tussles over virtually every feature in the park, including the presence of playgrounds and bathrooms, pedestrian access, benches and plantings and all the art work. These debates preceded the official opening of the park and are frequently revisited. Don Larson, a key activist in the CRAB Park campaign, and a CRAB-Water for Life Society organizer, claims that there is even a modest annual sum paid by the Vancouver Port Authority for a contribution to a Canada Day celebration and a kid's Christmas party "just to keep [him] off their backs." To concentrate solely on the monument's content would thus miss much of the intention behind it and its role in the life of the neighbourhood.

CRAB stands for "Create a Real Available Beach." CRAB Park is the name that local residents have used for years, rather than its more formal original name, "Portside Park." (As of 2008, the name was formally changed by the City to "CRAB Park at Portside Park.") The park's genesis was modest. As Don noted, it was "a discarded part of the environment of the city where everything wasn't tidied up perfectly." Some of the park's key supporters, in fact, laughed the first time they heard the idea

Reworking of CRAB logo from cartoon

of making it into a park, sometime in the early 1980s. Nevertheless, a few people began to piece together a campaign in 1982 – first a campaign of information, later one of involvement. A small office space was donated by Carnegie Centre. A series of letters went out to politicians and bureaucrats, and the activists conducted a survey of some 350 residents about what they wanted of such a park. University of British Columbia (UBC) students conducted an environmental inventory to detail and produce a report on the kinds of wildlife present. By this time the CRAB organizers had lost the Carnegie space and were operating their "office" out of a suitcase in various cafés. Soon they had even more of a problem: a bill from the UBC students for the publication of the report – one the CRAB organizers had no money to pay. CRAB Park organizers decided to hold a series of fundraisers, which had the effect of further focusing public attention on the project, as well as involving the organizers in a series of high-profile events. The group publicized its cause using the usual Downtown Eastside media – the Carnegie *Newsletter,* postings on telephone poles and word of mouth – as well as a series of CRAB-Water for Life Society mailings.

The first fundraiser – a festival on the eventual beach park site – took place in 1983. It drew not only local people, but also, somewhat unexpectedly by Don's account, two powerful federal politicians with "a huge frigging umbrella ... It was about ten by ten!" Perhaps the highly visible presence of the politicians kept watchful police at a distance from the crowd in a circular formation of paddy wagons. Simply being on the land was somewhat contentious; at the time, there was intense debate in advance of Expo '86 about a waterfront area so close to the downtown core. The lands were owned by the federal Port Authority but featured prominently in discussions at both provincial and municipal levels. It made a huge impression – not least on the surprised organizers – when the two politicians who attended that first fundraising event spoke in favour of the idea of making these lands into a park.

The organizers continued in their self-admittedly haphazard way, on a shoestring budget and with very little experience, to publicize the park through a series of events, festivals and street theatre performances. Though various politicians, residents, unions and churches had pledged support for the park project, the Port Authority abruptly announced it felt the idea unworkable. In reaction, in the early summer of 1984, two of the organizers started a spontaneous squat on the land overnight, and one thought to make and erect a pole, not carved with figures but, rather, adorned with the names of MPs, mayors, citizens and organizations that had spoken up in support of the park. The police didn't arrest the two organizers and, slowly, the occupation grew by word of mouth to include more than 60 tents. It continued for 75 days, and garnered enormous media attention. This was enough: word came from Ottawa that the area would, indeed, be proclaimed a park. After a series of improvements, including landscaping and amenities, finally, on July 29, 1987, the park was formally opened. Its legal status is provisional, however, in that the land is on a 45-year lease from the federal government. As Don Larson notes, "most of the acres are under the Park Board. Some of it ... is still under some kind of quote under the Port of Vancouver Corporation, after the lease is out. So these are dangerous things. Crazy things happen. They could ... they could build a high-rise there, just tear out the park." Indeed, the park is rather continuously threatened, both in terms of the land itself and in its aesthetic integrity, due to a succession of proposed development projects. To date, these have included a "Seaport Centre" project, a waterfront casino project and a soccer stadium complex. As it happens, the new Vancouver Convention Centre has just been built, at $900 million dollars, just to the northwest of the park.

The park is quite beautiful, and Don clearly goes there for retreat and solace, and tends to it personally, even though it is not his responsibility. On one December walk, this proprietary nature comes out (as recorded in my field notes): "Don walks me around CRAB Park ... Shows me his mother's bench 'dedicated before she died.' The waterfront is bustling, and strikingly beautiful. Cold, dry, water bright blue. A handful of people sitting, walking. 'It's always like this,' I mention, 'when I come.' 'Yes,' he says, 'except for in summer; up to 800 people for a concert then.'"

But all is not tranquil in the Downtown Eastside. During the same period that CRAB Park was becoming a reality, other things were going seriously wrong nearby. Women of the area were being murdered and disappearing at an alarming rate, especially since the early 1980s. As Marion Dean Dubik, an organizer with the Downtown Eastside Women's Centre protested: "There's people dying every day down here ... Look at the names on this list [over 90]. *All of these women counted.* We don't have Canada wide coverage of all the women that die right here in the Downtown Eastside. Why is that? ... We're not across the country. We're right here. We're dying, every day" (Morgan and Millar 1998).

But the violence in the Downtown Eastside is not just about dying, either. In an extended conversation I had with one elder, though we were supposed to be talking about a particular monument, she instead told me about her own life:

> She introduces herself as a Coast Salish woman – "my grandfather from Tofino, my grandmother from Sechelt." In the 1940s, largely to conform to the proscribed laws of the Indian Act, they [the woman's grandparents] married, though they did not often live together. Her mother grew up on Commercial and 7th [a residential area of Vancouver a few miles from the Downtown Eastside] ... Julia had seven brothers and four sisters. "I'm the result of a rape," she says, with a flat voice. And her life has been hard. She was married/involved with an addict for some 11 years, which leaves

you didn't bring yourself here
you won't take yourself away
even
if you stop your own heart
something remains
to be disposed

— Bud Osborn
little metaphysical poem

her little energy now, she says, for dealing with mentally ill people ... She adds "my daughter's dad was murdered." Julia claimed she herself has a history of heavy drinking, but now keeps herself sober and fit with a number of efforts: sewing and crocheting ("my grandmother taught me"), being in a poetry group, running, tai chi, meditation, and "having a lot of energy for other people."

... I am struck with how directly and frequently she speaks of needing to heed one's spirit to make decisions. Clearly – this is present for her – around her public role, around being with her daughter and granddaughter, around interacting with people on the street, around being a member of her community, being respectful, letting people know ... what she thinks ("it's not personal. It's my responsibility for they will think about it. Now, in a week, in three months. Sometime."), letting "patriarchal clan" people know her mind, spoken as a "matriarchal clan" woman.

When Julia laughs and says, "It's all been learned. It's all learning," I ask who her teachers are. Julia again replies with a story. She tells me that she herself had been very ill, mysteriously so from a medical standpoint, until quite recently. And then a relative had died, and then her unborn granddaughter had complications in the womb. It was after that, "the family began to understand." She explains, "It is our way to look at all the lives affected." There was a relative, she said, who was sick [involved in substance abuse], who nevertheless danced one of the traditional dances with a mask for that dance. Later, Julia was given a brooch of the same

> masked dancer. It was after that, she took sick – "this was the first sign – but I didn't understand it. So the illness moved on." Then the woman died, the fetus was troubled. Only then did they understand, and address the "root cause"; they instructed the man not to dance with the mask while addicted. Only then did the baby stabilize. "You see? Even as a fetus, this child has been teaching."

Upon reflection I realized something important but confusing: we hadn't talked at all about the monument. Instead, after many weeks of missed appointments, Julia had decided to speak to me, and then to invite me to speak with her again. Whether she intended it or not, my sense is that she did not so much tell me a story as bring me to a way of listening differently, a process that she only ever alluded to ("it's not personal. It's my responsibility for they will think about it. Now, in a week, in three months. Sometime.") For example, by simply narrating a version of her own life, Julia taught me, first, about the ways that violence, exclusion, legal frameworks and poverty had coursed through five generations of her own family. Her story was personal, but as the Carnegie *Newsletters* and other interviews showed me, it was also horribly normal for First Nations families in the Downtown Eastside. She also shared with me both secular and more spiritual ways of living with such damage – not only to speak out her own truth with dignity, but also to move between the dead and the living, the old and the young, the mundane and the ethereal in order to learn ways towards healing. These realizations would immeasurably change how I listened to the realities lived by the Downtown Eastside's residents, and also how I understood the ritual functions and everyday uses of the monuments. Not until two months later did Julia and I actually sit down and talk about the monument.

What did happen, however, was a series of exchanges over the next several weeks – some by e-mail, usually face to face – that resulted in an invitation that I come as a guest to a series of residents' meetings in the Downtown Eastside. This, in turn, enabled me to be included in a series of informal meetings over some six weeks with Julia and many others who knew the monuments and lived in the neighbourhood. I was able to listen and build on the record indicated by the Carnegie *Newsletter,* which began publication in 1986 and reflected an interesting evolution

> No jargon
> One speaker at a time
> Everyone's contribution is respected
> Speak only for yourself
> Resist the temptation to problem solve
> Learn and understand instead
> Create safety
>
> — from a flip chart drawn up at a Community Directions workshop

of the invisibility of women in the Downtown Eastside. For the first several years, the newsletters were notably silent about women except as objects of poetry or authors of pieces on housing. Occasionally there was mention of a missing woman. This began to change in the early 1990s. A vivid piece published May 15, 1991, from the point of view of a female rape victim, prompted one of the regular male contributors to write in the next issue a long response that criticized the original piece for unfairly targeting men and for not being specific. This in turn sparked an angry reply by an older, respected female frequent contributor who wrote that she had been raped and it had taken her 15 years to talk about it. "We will not have our pain denied," she stated. This sparked a series of pieces, primarily on violence and children, which began to link violence, drug addiction and the sex trade. The cover of the July 15, 1994, *Newsletter* was unambiguous: "Stop the Deaths! People are Dying from Overdoses and People are Being Murdered." There was a call for a community meeting. If the *Newsletter* can be seen as indicative of the neighbourhood's *gestalt,* it appears that, though murder and disappearance were certainly on people's minds, they weren't perceived at the time as part of a pattern.

In fact, the systematic and methodical nature of the women's disappearances was far more completely revealed in the search for one particular woman. In April 1998, Sarah de Vries disappeared. Her friend Wayne Leng put up posters asking for information about her, and asking that others who had lost family and friends contact him. He listed his own phone number. It was only over the next few months, as family after family telephoned, that the full horror of the violence against women in the Downtown Eastside came to light: 71 women were dead or missing in a mere 14 years, as compared to 2 women from 1900 to 1975 (Wood 1999, 102). Less than two years later that number had climbed to 144. Finally, police began to acknowledge that these disappearances and murders could not be coincidental. Criminologist John Lowman, who has specialized research expertise on issues of violence and the sex trade, notes that Vancouver's pattern shares no similarities with that of either Calgary or Toronto during the same periods, and suggests that multiple serial killers may well have been working the territory (Lowman, in Wood 1999, 102).

There is another pressing aspect to these women's (in)visibility that has to do with racism. Organizers for the CRAB Park boulder memorial are unequivocal: If these women had been white, their disappearances would have precipitated action (there *are* non-Aboriginal women who have been categorized among the missing, but they are in the minority). There is no way to prove this, yet there is a haunting detail. Sarah de Vries, whose disappearance prompted the telephone pole leaflets and the subsequent collection of details that resulted in the first collective confrontation with the police, was a black woman. But she had been adopted as a child into a wealthy, educated white family, and that family came forward forcefully into the public eye when she disappeared. Wayne Leng, the friend who began the search for her, is a white man. Is this mere coincidence? The question is especially pertinent given how deeply, and how often, colonialism has unleashed violence between British Columbia's First Peoples and its many immigrant populations.

Some of the key organizers for CRAB Park worked closely with people of the Downtown Eastside, and knew intimately of these deaths and disappearances; one, indeed, had lost 14 female members of his own family to violence (the Arrance family is mentioned in greater detail below). They also knew, from their work with CRAB Park, how strangely the media effected and affected public attention on the Downtown Eastside. Although one of the organizers claimed to have started the original Valentine's Day march in 1991, he knew that the marches had not succeeded in focusing more police or political attention on solving these cases. Some people interviewed claimed this was due to racism and (possibly) deliberate conspiracy in the Downtown Eastside itself. There was a sense of complete frustration with the police for not thoroughly investigating any of the cases – neither the murder cases nor the missing persons cases. And there was more: as Don said, "It's not just emotional rage. It leads to the right to control [people] politically, to control them psychologically. No woman can actually walk around the Downtown Eastside now without, consciously or unconsciously, being afraid. I mean, think about it. They put up a $100,000 reward for information about these cases, and no one claimed it. What's that say?"

Deciding that the lack of information and attention was intolerable, the CRAB Park organizers elected to do something "like a permanent newspaper article ... something they couldn't erase ... if you keep things in the public eye, they can't miss it." They hit upon the idea of a monument: "we need an angle, need a story, something a little bit new, for the media." The idea of a permanent and natural form appealed, and seemed in keeping with a First Nations' sensibility – an important consideration since such a large number of the women lost in the Downtown Eastside are from First Nations families. Moreover, boulders, like tombstones, are an "understood" form of memorial; Vancouver offers versions of boulder memorials in items as diverse as the Pauline Johnstone *Memorial* near Lost Lagoon, the somewhat odd *Girl in a Wet Suit* off the coast of Stanley Park, and a variety of Vancouver millennial public art installations.

The proponents of the CRAB Park boulder of course knew about the *Marker of Change*. Don Larson mentioned that he happened to be at the Park Board meeting at which *Marker of Change* was discussed in some detail. He was there on another matter, but remembered speaking up on behalf of the monument. He wrote to the Monument Project Committee informing them of his work on the CRAB Park boulder. It is hard to know exactly how the two affected each other, but there are similarities between how Don talks about the boulder and the promotional literature for *Marker of Change*. Don mentioned, for instance, that he and the CRAB-Water for Life Society "got an idea people need a place, need a visual reminder, so [the situation] is not so abstract. [The boulder] allows for small gatherings by being there. Maybe there's some healing ... [the idea of a boulder] is that it's permanent, natural. The idea is not to change it more than you have to. You know, there's a beauty to boulders. They're part of Scottish, Celtic, Irish, Japanese, Native cultures."

Perhaps because of their awareness of the tension surrounding *Marker of Change,* or perhaps because of a finer appreciation of how things get done (and not done) in the Downtown Eastside, the proponents of the boulder decided their project would follow a very different process: "[you know], the street people ... the down and outers ... still have absolutely no power at all in the community. That's the sad part. And they're so busy trying to survive, yeah there's drug and alcohol problems, well there's food problems and there's all kinds of problems and violence problems and homeless problems and all of that, everyday everyday everyday and that's what they're thinking about. They don't have a bunch of time to theoretically, philosophically think about your issue, whatever it is."

Don admits that he was "no good" at fundraising, that he doesn't like meetings, and that, anyway, people show up to one meeting and then often not again. He suggests that working to do things in the Downtown Eastside is complicated: "There's always going to be people saying 'I don't like it,' so you just have to do it." For a number of reasons, the boulder became more typical of most monuments in that it was commissioned, designed and paid for by a handful of people. It did not go through a juried art process, it did not attract negative press or spark a public outcry, and it did not require large amounts of money to be built.

The boulder itself is of modest design: an armchair-sized lump of rough granite provided by a funeral home. It was a direct private order. In contrast to the controversy precipitated by *Marker of Change,* only one letter was received by the Park Board regarding the boulder memorial, and it was highly supportive; there were no letters of opposition. After hearing testimony from the CRAB-Water for Life Society, the Park Board unanimously approved the application. The boulder sits in a naturalistic setting between a walking path and the beach. It was originally placed in this position so that it could be seen from Columbia Street and from the Vancouver Police Department steps; a Canada Customs building now obscures that view. One must now be in the park to come across the boulder, and even then, its presence is somewhat understated.

CRAB Park boulder | Note how the placement of the boulder, so close to the path and the water, make the monument a mixed site for interaction. | *Photo: Adrienne Burk*

The wording on the boulder, though it appears fairly simple, tells several tales:

THE HEART

In honour of the spirit of the people
murdered in the Downtown Eastside.
Many were women and many were native
aboriginal women. Many of these cases
remain unsolved. All my relations.

HAS ITS OWN MEMORY
JULY 29, 1997

CRAB Park boulder | The boulder sits very close to the central business district of Vancouver. Even the organizers were surprised at their success in securing such a waterfront site for the park, given developers' desires for the area. | *Photo: Greg Ehlers*

There are two accounts of where the line "the heart has its own memory" originated. One advocate suggests it was a personal poetic inspiration, the other said he'd read it on a Hallmark card and it struck him deeply. The rest of the inscription was the result of several decisions that speak to the complexities of the issues of missing women, respect for differing cultural traditions and simple logistics. For example, it was difficult to know for whom, exactly, this memorial should be designed. The inscription reads that it is "for the people murdered in the Downtown Eastside." This includes the victims of crimes over decades, women and men. A list of these names would possibly number in the hundreds. Consequently, "we didn't have enough room for all the names, and we didn't know them all." But the ambiguities were even more complex: In the absence of human remains and crime scenes, the police refused to classify those women reported missing by their families as murder victims.

The saga of the missing women is an iconic story, one that on close examination speaks volumes about the normative conditions under which lives must be lived in order to be regarded as of value and significance. These expectations have enormous influence on who is, at any one period, categorized as missing. The term "missing" is applied to people who embody identities as consumers, wage-earners and members of heteronormative family arrangements. At this point, authorities had – for nearly two decades – been unwilling or unable to comprehend that women who worked in the sex trade (and who thus might be temporarily relocated by pimps), women who often admitted to addiction and women whose financial and familial lives were thus disrupted could nevertheless live absolutely established patterns of existence. Families had tried for years to tell police that though, yes, their loved one was an addicted sex-trade worker, she nonetheless always phoned on a birthday, or always banked her welfare cheque or otherwise indicated a kind of structured life. These negotiations about legitimacy still continue as police visit and revisit files going back to the 1980s to determine who counts as missing. As of this writing, the numbers continue to change. They do so with depressing regularity, but they nearly always increase.

Nevertheless, no matter how certain the families were that a woman was missing, it would have seemed macabre to chisel in stone the name of someone for whose death there was no evidence. Further, there are various First Nations traditions around the use of names, which requires permission from elders in First Nations families. These traditions differ, as do the elders' decisions. However, among the oral traditions of many First Nations peoples, using a name can disturb the spirit, already suffering, or prematurely still what should be animate. In the end, the proponents decided to heed the advice of a city employee, to "stay with 'many.'" The final phrase, "All My Relations" is widely used in First Nations speeches, often as an ending to statements and prayers.

The monument is actively used in various ways. I have found flowers, tobacco, and small piles of stones, sometimes with names or wishes written on them, left at the boulder, especially since 2002. For "official" ceremonial use of the boulder, the key date is February 14 every year, when the boulder serves as a site for a family-organized ceremony. It has also been used as a site for interviews by various camera crews when yet another woman has been reported missing. In Spring 2009, it was the gathering site for a march from Vancouver to Terrace, physically linking the cases of Vancouver's missing women with those missing from northern British Columbia's Highway 16, also known as the Highway of Tears. I am unaware of any official dedication ceremony, although the date on the stone indicates July 29, 1997 (the tenth anniversary of the dedication of the park), and a *Newsletter* account of the celebrations that day mentions the unveiling of a "cairn."

There was, however, another highly significant use of the boulder in a ceremonial sense. On May 12, 1999, Sarah de Vries' thirtieth birthday, a memorial service was held at the First United Church in the heart of the Downtown Eastside. It was organized in large part by Sarah's sister Maggie de Vries, and Val Hughes, the sister of Kerry Koski, another missing woman. Several hundred people attended; I was crammed into a pew with many women who introduced themselves to me as Sarah's and Kerry's friends. The church was heavy with the scent of sweetgrass and candles, and we all took part in a long ceremony full of stories, prayers, and immense grief, a service that combined First Nations and Christian rituals. There was a smattering of local media, but this was more a service for a small community deeply scarred. Many spoke – family members, a local bus driver who remembered several of the women, individual police officers who were thanked (though police in general were criticized). Following the service, the mourners were joined by others and walked through the streets, drumming, singing and stopping traffic with photographs of all the women with the details of their dates of birth and disappearance, and a banner demanding "Find These Women Now." Eventually, the procession reached CRAB Park, where flowers, offerings and prayers were left at the boulder. As in keeping with traditional native ceremonies for thousands of years, these offerings included a scattering of ceremonial tobacco. In many Aboriginal traditions, tobacco is offered as a gesture of thanks, as a gesture of respect (before seeking advice, for example, from an elder) or as a gesture linking to the worlds of prayers, dreams or those who have died.

For this event, a small notice appeared in the local paper two days before (Service planned ... *Vancouver Sun* 1999). Curiously, it claimed that the services would end at CRAB Park but would also include the dedication of a memorial bench, though on the day I saw no such activity. Even more oddly, it wasn't until March 20, 2000, that another article appeared in the same paper announcing a gathering to officially dedicate a bench (Park memorializes ..., *Vancouver Sun* 2000). This article had a picture of people gathering around the bench that indeed now does appear several metres away from the boulder. On it is a plaque:

> In memory of L. Coombs, S. de Vries, M. Frey, J. Henry, H. Hallmark, A. Jardine, C. Knight, K. Koski, S. Lane, J. Murdock, D. Spence and all other women who are missing. With our love May 12, 1999

Thus the boulder, which was dedicated in 1997, appears to have functioned as a placeholder memorial form for an event nearly three years later, even though *that* event claimed a date nearly a year earlier than is recorded on its plaque. This memorial space, then, signals many things, but not a coalescence of clarity, or a settled memory.

In some ways, this monument did not follow but, rather, preceded its content – it has served to collect and focus previously unassembled griefs, confusions and anger. It has been subsequently revealed that the year of its dedication (1997) was the year that the number of missing women nearly doubled from previous years. One can only imagine the

"There is even a hierarchy of grief. Because of the need for sufficient DNA evidence to go to trial, only six families (of at least 69) could. Inside the logic of DNA evidence, you only matter if you left behind a hand bone or a hank of hair."

– Stevie Cameron, journalist

effect that disappearances had in the neighbourhood – one or two women every few weeks simply disappearing from the handful of streets where they were known as neighbours.

As in any small community dealing with trauma, there are several common responses. One of the most powerful responses is a drawing inward, a clearer and clearer separation from outsiders. It is perhaps not surprising, then, that for the December 6 event in 1999 (the 10-year anniversary of the Montréal Massacre), a piece in the *Vancouver Courier* announced:

> A coalition of Downtown Eastside women's groups called Breaking the Silence is organizing a walk from Pigeon Park [a street park on Hastings] to commemorate the Montréal Massacre, and the women who have died or gone missing on the Downtown Eastside. We see the connection between [the] violence that happened to the women in Montréal and the violence that happens to women every day down here ... the coalition *wanted to hold a vigil in the neighbourhood,* for women who would find it difficult to get to the art gallery. (Coppard 1999; my emphasis)

The neighbourhood defined in this quote does not include more than a few blocks. It is a neighbourhood measured in footsteps and known faces, a neighbourhood beyond camera lenses. It is a neighbourhood only six or eight blocks from the Vancouver Art Gallery in one direction and about twelve to Thornton Park in another, where other December 6 commemorations were being held. The fact that there was a separate ceremony in this neighbourhood, and that the location of the gathering was not the CRAB Park boulder, seems significant. It may have been simply that, in December, access to the boulder across the open ground towards the shoreline might have been difficult or perilous – it certainly would have been cold. But also, any gathering there would have been less visible: while Hastings Street, which runs through the Downtown Eastside, is a major traffic thoroughfare, the boulder (and the art gallery) is some distance away from the streets where the women disappeared and were murdered. Holding a ceremony in these places may have felt disjointed to Breaking the Silence. For example, the Valentine's Day marches make a point of visiting the precise locations where women's bodies were found, where women were last seen or where violence was done. These sites are as plebeian as a dumpster, a telephone pole, the doorstep of a hotel, the back of an apartment building.

The main ritual use of the boulder now is on the morning of February 14, when it features as a central presence in a family ceremony. On that day in 2002, about 40 people gathered on a sunny, cool morning, an eagle soaring overhead. The only way to have known about this ceremony was to have been invited or to have been very local; only a handful of notices were posted on telephone poles and bulletin boards in the area, all of which went up only days before. Tobacco was offered. Those present represented a range of races, ages, genders and cultures. One of the First Nations elders who attended remarked positively on this, saying that it takes diversity and the four colours of people to make a whole healing circle.

It was the Arrance family who had asked for the ceremony and who had prepared food for it. The family had been holding similar ceremonies for some years, having lost multiple female family members to violence. Fred Arrance, one of the brothers, has been consistently active in addressing the issue of the violence in the Downtown Eastside, claiming (with others) to have originated the Valentine's Day march, organizing for CRAB Park and the boulder, and serving for two years on the Carnegie Community Centre board, trying to generate support for a totem to be placed in CRAB Park. In addition, he has since contributed to other undertakings connected with some of Vancouver's millennial public art projects. Fred himself is from Prince George, and his heritage is mostly Cree/Métis. But he has lived in the Downtown Eastside for nearly five decades, which means he arrived as a young teenager. The Arrance family has been deeply active in the community, regularly cooking for feasting and ceremonial days in the neighbourhood. It was not surprising that elders from eight traditions came out to support them in February 2002.

A few days before, for the first time in 18 years, an arrest had been made in connection with the missing women. Robert Pickton, the owner of a pig farm in Port Coquitlam, BC, was in custody, and his farm had been seized as a crime scene. Emotion was running very high in the community, and the need for healing was palpable. As the ceremony began, an elder made an offering of tobacco to the sea as thanks "for the blessing of the eagle" soaring overhead. After this, the group was asked to make a large circle, arranged around the boulder as if it were another person in the circle. The elders prayed, sang, drummed, smudged and spoke in several languages, each taking a turn in the centre. One of the elders welcomed us and said, "Things always change, as they do, but we are here at the ... Aboriginal memorial" to do honour, to give comfort. This was the first time I had ever heard the boulder referred to in this way. Many times the Arrance family was thanked, many times they were honoured for their big hearts and respect. The rhetorical cadences used in the ceremony were striking. One female speaker in particular never spoke in complete sentences, but would, through a juxtaposition of lines, linked often by the phrase "in the highest honour," suggest many possible directions for interpretations. Her words thus simultaneously served as a sort of invocation of the divine, an accusation of the secular, an assertion of the community present, both corporeally and in spirit, and a running commentary.

It was evident from the clothing worn by some of the elders and the presence of drumming that we were engaged in a ceremony, but this did not stop one white man on an apparent power walk from charging through the middle of the circle as an elder was speaking. He was asked, "Where is your respect?" to which he screamed, "Get over on the grass! This is a public path!" A Navajo media team (which had permission to record the speakers with microphones but who stayed outside the circle and at a distance in order to photograph the proceedings) caught the man's callousness and arrogance on tape, but there was no physical attempt made to remove the intruder.

This may have been one of the only encounters with the media, other than at an official press conference, at the boulder. And as with *Marker of Change*, the effects of the media were complex. Don Larson mused, "It's a mixed thing about media. Without it, you're dead in the water. But with it, it's complicated." Don generally doesn't like the media, partly because, in his words, "I don't trust what I'd say in front of a camera" but also because "no one needs a camera in their face when they're mourning." In any event, there was no other media present at the family ceremony, and it seemed clear that permission had been sought and negotiated carefully with each of the elders and the family for the Navajo team to be there. Permission is important in the First Nations tradition; as an elder in another ceremony reminded us, "songs are like our constitution in our traditions." Elders need to protect not only the texts of their traditions – songs, prayers, stories, masks, dances – but also the circumstances of their display and circulation. On that February morning, the prayers continued until we closed the circle by shaking hands with the family, who thanked us all for being there. We wished each other a Happy Valentine's Day, to remind ourselves that the day is traditionally one for celebrating love.

In the quiet serenity of the boulder in daylight, I have observed at most one or two people near it, standing or sitting on the beach, or perhaps on the bench. Offerings left at the site are either infrequent or indiscernible (since they may blend in with the natural elements surrounding the boulder). But these facts should not be misread; parks are important imaginatively as well as physically, and certainly this park is claimed bodily in the summer months, whether or not that claim is enacted in the colder and wetter months. In any event, of the three monuments discussed in this book, the boulder's placement is the most conducive to reflection and even meditation. And many Downtown Eastside activists and residents have commented on how few such spaces exist. Remarked one individual, "We don't have flowers down here. There aren't enough invitations to stop. Life moves pretty fast down here."

Standing with Courage, Strength and Pride

Given that both the boulder and *Marker of Change* were already in place, the impetus for the third monument – a carved pole in Oppenheimer Park – appears puzzling. Though of the three monuments the reasons for the pole are the most obscured (by conflicting narratives and comments by some of those interviewed that implied some responses may simply have been made up), one elder explained to me that, despite the existence of the other monuments, there was a need for a monument "done *by* us, not *for* us," which is why he became involved. But who is "us"?

One answer is the current First Nations residents. Oppenheimer Park, like the other two parks, lies on lands claimed by three First Nations: the Burrard, the Squamish, and the Musqueam. Many members of these nations and others (including other First Nations and the Chinese, Latin American, African, Japanese, and European communities), live either literally in the park or in densely occupied hotels, houses and apartments nearby. Thus, on a daily basis, those inhabiting the park are representative of a truly multinational population. But Oppenheimer also exists within a more recent geographical expression, that of the city grid. Of the three parks considered in this volume, Oppenheimer is the smallest and is situated between the other two – in the centre, as it were, rather than on the edges of the Downtown Eastside. It also occupies a different imagined location as a central feature of a changing area of the city: "Through the 1890s, the area experienced rapid development of housing for both wealthy and labourers alike. By the end of the decade, however, the wealthy had relocated to the West End and Fairview Slopes, leaving the Powell area distinctly working class. The extension of the street car line on Powell Street in 1890 also spurred commercial development – initially in the 200 and 400 blocks – transforming these from houses to commercial and hotel/boarding house uses."[15]

Oppenheimer became a formal park in 1902. Known then as the Powell Street Grounds, the park began as a village square and over the past century has been the site for circuses, key labour demonstrations and bloody confrontations during the Depression between police and the unemployed. It has also been, for nearly 60 years, the site of a much-loved and much-used baseball diamond. At various times during the 1970s to 1990s, the park received (relatively) substantial sums of money for improvements and amenities. Today it houses a community outreach

15 http://vancouver.ca/commsvcs/planning/dtes/oppenheimer.htm.

building (linked to Carnegie), a kitchen, bathrooms, children's play areas, numerous benches, art pieces and ceremonial trees, a horseshoe pitch and checkerboard, chess, and shuffleboard areas. It is an intensely used park – I have never seen fewer than a few dozen people in it, day or night. As one person put it, "I live in this park ... I sleep in a room, but I *live* in this park." Indeed, several Downtown Eastside residents in casual conversations characterize it as either the neighbourhood's "backyard" or its "living room."

In such a diverse community, such a communal/intimate space is the location of an array of activities and human drama. The realities of homelessness mean that much of this is on display. Steve Johnson, who was employed as the Carnegie Community Centre outreach worker during the carving of the pole, explained the daily realities of life in Oppenheimer by first discussing his own introduction to the neighbourhood and how to work with its residents:

> When I first started there [Carnegie], I was very green. I was thinking one day that people don't decide to be homeless, it just happens to them. So I decided one day at work just not to go home, to just suddenly be homeless, to see how it was. I spent the weekend with a woman named Shampoo. She was an older woman – she's died now – that lived on the streets for years. I spent two nights out. I slept under the Georgia Street viaduct, and people threw bottles at us, harassed us just for being there. It was really cold and windy. But Shampoo really impressed me. When we met some kid – no matter what shape they were in – she'd listen to their troubles or whatever thing they were in, and then she'd make them agree to meet her again. She'd say, "you have to check in with me." And the kids would do it. I realized then, this is how to do this. I decided I wasn't going to do traditional social work after that.

But Steve learned that life in the Downtown Eastside took enormous tolls. Much local legitimacy is claimed in terms of familiarity with hardship and with embodied experiences of violence and addiction (that is just what Julia had suggested and what the residents' meetings inscribed

> "If more people were screaming, I could relax."
>
> — Passerby on Hastings Street discussing the missing women, overheard by the author

over and over). Steve used to warn Carnegie volunteers (who often came from other parts of the city): "I [explained] to the new recruits every summer ... You know, there are about 10,000 people down here. You'll meet about 500-600. You'll affect about 10, maybe 15. And of that number, at least 2 will be dead by Christmas." Such statistics inevitably have a way of manifesting in mundane ways. At the Oppenheimer outreach office, Steve said that the morning routine, every day, took about two hours. He and a co-worker systematically went around the perimeter of the city-block–sized park, cleaning up the grounds, shaking people awake, "basically checking to see that no one was dead."

When he began working at Oppenheimer, Steve remembers, there was a great deal of drug dealing and drug use at the park. And there was also the high drama that often accompanies both activities. He recounted one incident, for instance, when a junkie was in the throes of a bad high in the centre of the park, and the police response, complete with wailing sirens, resulted in the junkie stabbing himself, cutting an artery and bleeding profusely. In another case, Steve was shot at by drug dealers after he took to openly videotaping them. Over the years there have been many initiatives, led by different people – in attempts to secure the park as a drug-free space for kids, as a safe place for the elderly and as a haven for anyone needing refuge from the squalor of a single-occupancy hotel room. Indeed, the park is ringed by soup kitchens, missions, community halls, churches, some small businesses and residential homes, so there is a ready constituency of neighbours who immediately benefit from stability and non-violence in the park.

These diverse groups of people, then, in some way constitute "us" – bound by kinship ties that are partly geographical, partly cultural and partly experiential. According to Steve, the idea for a pole came out of an awareness of these eclectic but strong bonds. Following a funeral for a friend who had died from drinking, several individuals decided to sit in the park and talk about their deceased friend. Steve said that what struck him at the time was that they all sat down "and jumped right back into drinking." He thinks it was Shampoo who said, "You know, no one knows that this is their place. This is their home. This is their house. We need to *show* that this is our house." I have heard another story, too, that there was actually a permit in place in the late '90s for a pole in CRAB Park, but it had never been erected. Whether or not another pole project had been previously conceived, Steve claims that at the time, he was in a position to get Carnegie to support the idea.

Of course, poles are quite a common feature of the Vancouver landscape, from the collection of poles from many traditions in Stanley Park (especially popular with tourists), to a collection of poles more fully contextualized in the University of British Columbia's Museum of Anthropology, to the carving and raising of poles for a variety of commercial and secular enterprises. Therefore, like the idea of the CRAB Park boulder, the idea of a pole was to use a familiar form as a public art piece.

Traditionally, choosing a log for carving the pole is a highly significant event. It can take months, since the pole must be of sufficient size, possess straight grain and have only minimal rot. Along the northern Northwest Coast, a family would commission a carver from another family to carve the pole, which might include the task of finding the log; alternatively, the log might be a gift. In the case of *Standing with Courage, Strength and Pride,* traditions were impossible to follow: there was no single family to commission the pole, nor, according to Steve and all the other people interviewed, was the pole ever meant to be exclusively for First Nations people. Instead, the pole was to be for "everyone who's died in the Downtown Eastside." According to Steve, there were initially a few attempts at securing a log. Though there are no recorded details about this, Steve claims the first pole came from Seaton Lake, but was determined to be too small. At that point, Tsleil-Watuth Chief Leonard George told Steve he had two logs from Rivers Inlet Burrard Reserve, and donated one to the Oppenheimer pole project. Originally, it had been intended for a canoe, but it had been dropped and had subsequently split. The log arrived over Labour Day weekend, 1997, and was placed in Oppenheimer Park. It arrived, in other words, within six weeks of the boulder's installation and the groundbreaking for *Marker of Change*.

After the log is selected, traditionally there is much made of how the log is cut, which in turn determines how deep into the wood a design can be carved, the extent of the carved area and, thus, what kinds of figures and intersections of figures can appear. Steve recalls that the donated log had extensive rot, that it had to be inverted (so it would be uncharacteristically thin at the base) and that it had to be deeply hollowed out. According to Steve, work began immediately, including hollowing out and rounding the pole, planing off the rough bark, even imagining the large figures. Steve's idea was that anyone who wanted should be able to work on the pole: "My goal was that I was not going to lose sight of the people ... anyone who contributes to our family in the park can contribute to this pole." Steve claimed to have done some carving, but admitted he had never worked on something as large as a pole. The design work was given to another young man, but he, too, was quite inexperienced; though he drew a design, once the log was in front of him, Steve said the young man realized he "couldn't work that big. He kind of freaked out."

In November of 1997, a noted carver, Dick Baker Sr., came through the park to see what was going on. According to Baker, "[Steve admitted] they were stuck. He asked me to help. And in our tradition, if a Native asks you, you cannot refuse." Dick was not pleased with how the log had been cut. He said the extent of the hollowing out was unnecessary, and gave them "less room for mistakes." But Dick, too, felt the power of the idea: "the pole was needed for a long time ... We didn't want to build a monument, a memorial. We didn't do it for the building of it, for the money. We did it because people needed a place to come. Some of these guys have no other place. A place to be still. For peace to come. They can look and say 'I did that.'"

Dick was known as a master carver. He had been carving since he was nine years old, training with Ellen Neel, who worked in a shed in Stanley Park beneath what is now the Tea House. His own life experiences exposed him to many Native traditions ("I'm from the Squamish Nation, but I've known all kinds of people. I married into Alert Bay people. And I'm a fisherman, so I've gone up and down the coast"). "Master carver" is a term that has changed in meaning in more recent times, but to the Nisga'a artist Norman Tait, master carvers share certain characteristics:

> In my opinion, a master carver ought to be able to design and execute anything – masks, jewelry, flat design, totem poles, canoes, houses. I also think a master carver ought to be able to work in a variety of media – wood, silver, bone, flat design or bark. Furthermore, a master carver ought to be experienced enough and knowledgeable enough to be able to carve in a variety of styles, not just the tradition of the place he comes from. He should understand the differences in style and also know how to achieve them. (Jensen 1992, 10)

Dick looked at the design provided and found that it would not work well for the log. In traditional carving, the design of a pole is guided only minimally, because the pole is meant to manifest experiences that people already know. But such traditions are changing:

> The crests or figures on a totem pole generally relate to the identity of the owner of the pole. Often it is assumed that knowledgeable people can "read" and understand a totem pole merely by identifying these figures. In fact, the combination of figures serves more as a memory prompt for those who already know the identity and stories of the pole's owner. People who witness the raising of the pole and hear its story proclaimed are expected to remember both the occasion and the narrative whenever they see the pole ... [traditionally] the owner of the pole would tell the carver which crests to put on the pole but would usually leave the design up to the carver. Today, totem poles are commissioned by groups ... [without] traditional family affiliations or crests. So customs and traditions have to be reshaped, and in such instances it has become common for an artist to design the pole around a story belonging to his own lineage or one that he has been given permission to use. (Jensen 1992, 13)

Dick's four decades of carving and his exposure to many traditions fit well within this newer approach: he had permission to carve certain

Standing with Courage, Strength and Pride | The senior carving team for *Standing with Courage, Strength and Pride* was made up of experienced carvers from several First Nations, but residents and neighbours also contributed to the making of the pole in many other ways. | *Photo: Steve Johnson*

figures, but also knew about other traditions. Moreover, he assembled an experienced group of carvers ("guys I asked on the Capilano Reserve") who were indeed diverse. Paul Auger came from Athabaska, Alex Weir from Alert Bay, Kim Washburn from Haida Gwaii. Dallas Hunt, a Kwakwaka'wakw, came as well, even though he usually worked now only on masks.

When asked about the redesign, and how the represented figures fit within this Downtown Eastside neighbourhood of so many nations and so many traditions, Dick replied, "But these animals are known all across the country, in all traditions – except for the *sisiutl,* which is Kwakiutl – but it is known widely as the most powerful creature on the coast. If you see it, it is an omen. If you touch it, it is death. Raven is a messenger. So are the wolves. The wolves between mountains." He reiterated to me that such knowledge is important, as one does not write down the story of a pole.

When the carving began in earnest, the work was focused and intense: nine o'clock until two o'clock, seven days a week, for nearly five months. And as was the intention, this monument really was done *by*, not *for*, the community. Every day, the carving of the pole took place in Oppenheimer Park, with one exception. Work on the Raven figure did take place in other locations. Again, accounts vary, but it appears that there was a large separate piece of wood for the Raven figure, but it was difficult to "see" the figure in the wood. One day a gifted carver from Vancouver Island (none of those interviewed could remember his name – "we call him the Lone Ranger") happened by the parking lot where several were looking at the wood, without the confidence to cut into it. This mysterious carver saw the design for the figure, asked for permission and was given the chainsaw, and roughed out the figure in moments, leaving several astonished viewers. Others worked on laminating the wood before the Raven was placed at the top of the pole. Other than the Raven, all the carving was done in the park.

Activity in the park took many forms. Often, an eagle or a pair of eagles would sit in one of the park trees, or several would soar on the thermals high above. Because eagles are the creatures that can fly higher and see farther than any other, they are considered powerful signs of the "spirit world" by many Aboriginal (and non-Aboriginal) peoples. The task-related activity in the park was concentrated around the team of five main carvers. Others in the community could help with the rough work, and with small efforts and gestures that are also part of the work of the pole, including rolling the pole in and out of the carving shed.

The carvers were experienced, and had been taught in traditional ways. Dick brought his own tools, while Paul Auger made his own tools – often a key part of working on a new pole. The carvers knew each other, and knew the discipline of such large work, though they had not worked together on a pole before. Of course, the carving of a pole is always largely a social and even a public event. So in that sense, working in this place may have seemed in some ways quite familiar to the carvers, even though the setting was an urban one rather than a reserve or a studio. But after all, this *was* the Downtown Eastside: during the first few days, Dick said, there was a lot of pot smoking, drugs and drinking. After three days, he put up a sign: "Drink/Drug Free Zone."

There are conflicting accounts of how addictions were handled during the five months of carving. Steve Johnson remembers that what he felt most strongly about was the process of the pole carving rather than the outcome. He tried to ensure that every time people came by to work on the pole, they got the chance to do so: "A lot of people didn't know how ... but they came and worked a few hours. Sometimes much longer. One family, who had just lost someone, came for three days. Three generations." He noted that many people knew the traditional proscriptions that those battling addiction should not pick up carving tools. Steve recalls, "Some wouldn't touch a knife – they knew our culture. One guy said 'I'm still practising' [I'm using drugs]. But every day he came and swept up cedar shavings and gave them to people. But he knew, when drinking or addicted, he shouldn't carve." Others came and just watched. Steve said what was most amazing to him were the twelve people who came and worked but knew that, because of how sick they were, they wouldn't live to be there at the pole raising: "Between their addictions and AIDS ... they knew they wouldn't make it." Several wrote letters that said, "I helped carve this," and gave them to Steve with the instruction, "Don't give this to my family until I've died." This moved Steve, particularly given what he had observed of families who, in his opinion, often abandoned their own. "All their families say the same thing – and it pisses me off. 'If they'd just gotten sober' or whatever, they would have been welcomed back." Some of the carvers themselves did not live to see the pole feasted.

For his part, Dick knew the reality of life's pressures and the availability of drugs and alcohol in the neighbourhood. He identifies himself as someone who had been an alcoholic for many years. According to Dick, the carving would stop for one hour at lunch time and the five main carvers would walk to attend an Alcoholics Anonymous meeting at the Dugout, a community gathering place a few blocks away.

The need to be sober when using carving tools is only one of many traditions regarding how respect is to be shown during the carving of a pole. Others include the offering of food to the carvers and the ways in which visits by elders are handled. The offering of tobacco is important, too, as well as more prosaic support, such as, in this case, praise from the Carnegie staff and money for bus tickets. Accounts differ sharply as to

Raising *Standing with Courage, Strength and Pride* | *Photo: Steve Johnson*

how well these traditions were observed, and there is considerable criticism of Carnegie's role in these affairs. And the comments went both ways: One Carnegie employee claimed the whole process was "a bit of a nightmare" and that the tears shed at the raising were as much from relief as from pride. Others say the abuse was overwhelming; Steve acknowledged that one of the carvers was "one of the most difficult people I have ever worked with." For his part, Dick claims that at one point tension became so great that he "went to Carnegie and demanded that the staff come and tell these guys they were doing a good job even though they were from skid row. I brought in a crane to remove the pole and leave them a dead log instead to show them I meant it."

The protocols regarding respect even extended to the ceremony of the pole raising, on June 6, 1998. There was some difficulty raising the pole, Dick said, because "X claimed to know how to tie knots, but he didn't. Paul was halfway up the pole before we caught on." Traditionally, it is also at the pole raising that the narrative of the pole is told and witnessed, and that the carvers, who have honoured the community with their work, are in turn honoured by the community, with ceremony, singing, dancing and feasting. Here the accounts conflict again. On the one hand, a programme from the Carnegie archives identifies a ceremony marked by both Native and non-Native speakers and music, prayer and dance from different First Nations. The programme also contains a short statement (by Steve Johnson) about the pole: "Anyone who has spent time in the Downtown Eastside knows that hardship is a part of daily life. Good people die violent deaths here – from drugs, rice wine, diseases like HIV and Hepatitis, or at the hands of a stranger or friend ... This pole is not only a memorial to our brothers and sisters who have died unnecessarily in the Downtown Eastside. It is also for those who have survived and continue to live in this neighbourhood."

There is also a full page of thanks to individuals for particular tasks, to those who financed the project and to those who donated services. A poem written on the occasion suggests a significant community event:

> So did the First Nations people,
> With their friends and allies,
> Raise a totem pole
> In Oppenheimer Park
> On June 6, 1998
> to remember the community
> of those who have died
> in the Downtown Eastside
> and so did they rededicate themselves
> to the struggle
> for hope and for justice
> from one generation to another. (Cameron 2000, 70)

According to three individuals interviewed, however, though they claim that approximately 3,000 people were present at the event, no singing, no dancing, no ceremony and no feasting occurred. There was no media, "because of the disrespect of no feeding." When asked why they thought this had happened, one of the three said, "Because someone was

Feasting *Standing with Courage, Strength and Pride*, March 2001 | *Photo: Adrienne Burk*

doing it on the cheap." In any event, an acceptable feasting event occurred, finally, in March of 2001. The feast was prepared by Fred Arrance's family – the same family that cooked for and convened the ceremony at the boulder. There was also another programme for this occasion, and in this one, the pole is named: *Standing with Courage, Strength and Pride*. This time, there were singing, dancing, prayers, ceremonies, gifts, the telling of the story of the pole and the blessing of the food.

The "interpreted story" that appears on this second feasting day programme is significant for at least two reasons. First, it does not actually mention all of the figures on the pole. Second, such an account is precisely what others claim "traditionally" should not be written. Perhaps for that reason it is introduced as an "interpreted story of the pole[,] a direct quote from Steven Johnson":

> At the base of the pole are three copperheads that represent the people of the West Coast.
>
> Double Headed Sea Serpent which is representative of the good and bad in all things.
>
> The Sitting down figure ... [to myself] represents the inaction of agencies, different bureaucracies, and the helplessness we feel in the Downtown Eastside. It also represents a challenge for the people of the Downtown Eastside to not be victims of whatever or whoever ...
>
> The Mother with Child represents the generations lost here.
>
> The two Wolves represent all the people across Canada.
>
> The Raven represents the Creator without whom we all would not be here.

Although in a way Steve was present on the feasting day in the guise of this written statement, he was at that time living far north of Vancouver in Nisga'a territory, and it was clear he would not be at the ceremony in person. At the First Nations Caucus Working Group planning the event, to which I was invited, money was discussed for numerous items: food and tobacco and the groups who would do music, and possibly for renting a tent; no mention was made of money to bring Steve to town.

In any event, in some senses this day "finished" the welcome of the pole into the community. Indeed, many say that it is when a pole is feasted that it becomes a living member of the life of a community. On this second ceremonial day, too, a plaque prepared by Dick and his son was revealed. On it is this inscription (all spelling is as it appears on the plaque):

> Memorial Pole
> To our sisters and brothers
> who have died unnecessarly in
> the downtown eastside and to those
> who have survived.
>
> Many thanks to the exceptional community of volunteers and carvers of the totem pole. (carvers) Richard Baker Sr., Delbert Weir, Kim Washburn, Dallas Hunt, Luis Joseph, Maynard John, B.C. Matilipe, Matthew

> Baker and David Laird. Donations from Alaska Copper, frazer river pile drivirs LTD, cement – Tim Bestland and ABD crane LTD – Don Dadey.
>
> Figures on the totem top to bottom. Raven, wolves, woman and child, *sisiutl,* moon, wolf, whale, bear, and coppers.

Even though the plaque was prepared for that day, it was not mounted, even over a year later. Until September 2002, it sat in an office in the Carnegie Community Centre. When I asked about this, I was told that "authorities" did not want to put up a plaque with spelling errors. It was quietly erected in the following season.

Since Installation ...

Of all the monuments considered here, the pole is by far the most used. It sits, as mentioned, amid all sorts of activities in Oppenheimer Park, and items such as candles, stones with names, tobacco, small messages, signs, ribbons and flowers are left in personal gestures of memory and honour. The pole's first publicized use came within a month of its raising: on July 5, 1998, five elders began a fast in nearby Pigeon Park "for equal rights for all and for recognition of residential school consequences." They fasted for four days, then walked to Oppenheimer Park, where they gathered around the pole. The pole functions similarly every February 14 as the central figure of the closing ceremony for the Valentine's Day march.

As a developing tradition over the past several years, the first event on February 14 is held at the CRAB Park boulder; after that, it is possible to take part in events at various locations. On many years, there has been a little food and flowers left as offerings near the boulder at a green space adjacent to CRAB Park, another park that Don Larson was instrumental in establishing. Named for Wendy Poole, near the site where she was found murdered, this park is the first in British Columbia named for a First Nations woman. Many of the elders who attend these events may leave to have lunch in a nearby café. At Carnegie, two blocks away, an hour-long memorial service features words from families and friends who have been touched by or are active concerning the issue of the missing women. The ceremony there is respectful to those who need to speak

The route for the Valentine's Day march changes each year according to the wishes of the families and the increasing number of disappearances. | *Photo: Adrienne Burk*

for as long as they need to speak, but is also programmed fairly tightly, including an opening sage ceremony, prayers, songs and a closing thank you. The crowds who come appear drawn from across the city – it seems safe to say that some of the people who come to this event may come to no other events in the Downtown Eastside all year. Before the ceremonies end, it is customary for one of the speakers to remind us of the spirit of the day. In 2001, the mother of one First Nations woman horribly murdered implored us, smiling, "Today, put your anger somewhere else. During the march, [the missing] can touch us if we walk without anger." Just after this, the Carnegie Theatre empties onto the streets, where hundreds of people encircle the intersection of Main and Hastings, stopping

Encircling Main and Hastings at the Valentine's Day march, 2009. | *Photo: Adrienne Burk*

traffic, drumming, singing, and holding hands, banners and now quilted panels, each panel representing a missing woman whose family has given its permission to mourn her publicly.

The Valentine's Day march, now run through the auspices of Carnegie Community Centre but with coordination and leadership from some two dozen organizations and several individuals, began nearly two decades ago; its origins are now somewhat contested. It is now a considerable and growing annual event. (Indeed, the 2009 February 14th Women's Memorial March Committee claims there are now companion events on February 14 in Toronto, Edmonton, Calgary, Winnipeg and Victoria; the 2009 Vancouver march, advertised for the first time ever on Facebook, drew close to 2,000 marchers.) It is also an intensely emotional experience. As an individual who has been asked to emcee the Carnegie ceremony over many years told me, "I can emcee indoors, but not outside – I'm too vulnerable to the tension of spirits and hurt feelings." She claims the feelings she experiences during the event are

The banners name women whose families have given their permission for the names to be used. | *Photo: Adrienne Burk*

complex – "it's the *community* that does the event, but only certain people – Breaking the Silence, Carnegie Centre, the Downtown Eastside Women's Centre – come forward." In the years since I first attended the march, though the route is never exactly the same, the rhythms of activities have stabilized to feature certain events. First, after the ceremony at Carnegie, participants come out to encircle the main intersection of Hastings and Main, unfurling yards and yards of banners naming the women who have been found murdered or who have gone missing. This activity stops traffic in all four directions for about a half-hour. Then the crowd follows female elders on a slow march through the Downtown Eastside, with pauses to allow for offerings of prayers, tobacco and roses at numerous sites. Eventually, the march reaches the steps of the Vancouver Police Department, where a rally/confrontation is held, then carries on to Oppenheimer Park, and finally ends with a feast at the Japanese Language Hall. The whole event (starting with the boulder ceremony at 10:30 in the morning) lasts until twilight.

Leadership for the march was advertised in the programme of 2002: "The February 14th Women's Memorial March is organized and led by women because women, especially Aboriginal women, face physical, mental, emotional and spiritual violence on a daily basis. We invite the whole community to join us in the spirit of the march. We ask that men share their grief and show their support by respecting the structure of the march."

The role of men in the march appears to be changing, however. While I was on this march in 2000 (there were only about 100 of us that day), I was accosted by a man who asked if I didn't think the women "had it coming ... sort of asked for it, I mean." Later, when we were conducting a ceremony at Pigeon Park, a very drunk man made a point of screaming at the top of his lungs, just at the moment silence was asked for, "What is this, a lynching?" When told it was a memorial for women who died, he replied, "That's what I said. A lynching." A few years later, men played a very different part, as this account from my field notes shows:

> One man, walking beside me, sees someone standing on the sidelines. We are two in a large crowd moving very slowly down Hastings Street, blocking traffic, and summoning, with our drums and our singing, our banners and photographs of the missing women, more hundreds who watch the march while perched on window ledges and in doorways. My neighbour is so soft-spoken I lean in to listen. "Yeah, that's the guy there. You know, he's killed sixteen women with alcohol poisoning. But here he is, out on the street again." We walk by. I am amazed at the intimacy of the violence here – another, next to me, says, quite matter of factly, "The ones who murdered my daughter. I know who they are. They don't know who I am. I walk by them every day."
>
> We are on these streets – a march led by First Nations women elders, but a procession full of men and children as well – to stop at precisely the spots where women's murdered bodies have been found, or where they were last seen alive. And yet, one mother has

> implored us to walk with love, not anger. And so this is an odd march, drawing women who would never otherwise come to this neighbourhood to walk side by side with the acutely bereaved. Strange too because the locations, though they vary a little each year, keeping in grim step with the crimes and the permissions of the families – are in the heart of a row of tourist shops, along one of the main east-west streets of the city, at dumpsters. At each location, there is silence, prayers are offered, some ceremonial tobacco scattered.
>
> When we arrive at Oppenheimer Park, we are instructed to surround the pole at a distance, so that we can see each other, with the pole in our midst. People speak, but the circle is too large to hear. I look instead into the skies that rise over the sidings of the old houses and the soup kitchens that surround this block. I notice one – two – then finally up to a dozen soaring eagles. I know that one pair nests nearby, even in these noisy streets. They loop and fall like black notes against the high clouds, and I try to imagine what we look like from their sharp eyes.
>
> Something has been said; the circle shifts. A speaker comes near enough to be heard. She asks for those to have lost someone to these streets to move inside, form an inner circle closer to the pole. Until that moment, I realize I have been thinking this is all about adults. But the twelve year old appears – daughter of a disappeared woman. Then a six year old. Some toddlers. Grandmothers. Brothers, I imagine. Soon, over sixty people. The hush crushes the heart.

Though the role of men appears to be changing, some things that reveal the structural violence within the neighbourhood do not. I realize this as I observe:

> This gathering is not only about death. As I look around both the inner and outer circles, I see it is also about surviving against the odds. I think of Julia's five generations of violence. That is how it is, I realize, when violence isn't an occasional visitor, but moves in and settles into your life.

"We look at the destruction around us and perceive our collective poverty. We see that everything that is truly needed by the world is too large for individuals to give. We find we have only ourselves. Our experience. Our dreams. Our simple art. Our memories of better ways. Our knowledge that the world cannot be healed in the abstract. That healing begins where the wound was made."

— Alice Walker, *The Way Forward Is with a Broken Heart*

I watch the group cluster in and around the pole. I think about how the housed, the waged, the befriended, the healthy, and the un-assaulted in society can lean into social supports when violence strikes. But such supports apply unevenly to those gathering before me. In a kind of amplified cruelty, these people have lost, it seems, doubly, in life chances. They have reaped from life both more violence and less social support.

And yet, this pole proclaims, this community is made up of not just victims, not just the dead. The pole's figures, carved in rich, swelling strokes reaching towards the sky, lay that claim: the three copper shields of the nations, the signature clans (bear, whale, wolf), the moon and the *sisiutl,* the woman and child, representing generations, the wolves howling, the watching raven. "We survive. We are still here. We stand with courage, strength and pride." I notice the cameras are still present, but something else is going on here too. This is a place framed in the lens of a camera or the cast of a journalist's phrasing, but, more deeply, it is also a place from which to look back, frame the framers, frame the act of framing itself and say "We are more than you see. Look at your need to border and label us, and then perhaps you can begin to see us."

For seeing, and being seen, is an issue of great sensitivity in the Downtown Eastside. As noted by many, it is a neighbourhood overwhelmed by the effects of visual framing. According to several local residents, news media have been known to actually pay addicts to shoot up for their cameras, and frequently drive vans through the area with cameras rolling,

These large sidewalk mosaics were part of the "Footprints" community art project, which, like the pole project, allowed people to participate as they were able and wished to, under the guidance of experienced artists. There are several dozen of these mosaics throughout the Downtown Eastside. | *Photo: Adrienne Burk*

as one activist noted, "filming like we're in a zoo." The local police have formed their own video squad to use video as a community tool. One commercial photographer, Lincoln Clarkes, controversially photographed a number of prostitutes for ends that were not at all clear, financially or artistically. A renowned photographer, Stan Douglas, who for years maintained a photo studio in the neighbourhood, produced a book about the 100 block of Hastings Street, in the core of the Downtown Eastside. Perhaps it was inevitable, then, that the Valentine's Day marches would be fully framed in the glare from the media spotlights. Although cameras have traditionally not been permitted in the Carnegie ceremony (although the 2009 event was different), hundreds of photographs and videos are taken every year all along the march route; reporters from print dailies and weeklies, sometimes even entire film crews for local TV stations appear amid the marchers, complete with fuzzy microphones and heavy cameras.

But the pole somehow mutes this clamour with its own physical presence. Unlike the other monuments, the pole is tall, reaching for the sky. One cannot contemplate it in full except from a distance. Furthermore, the pole is placed right up against a walkway – where one can neither sit nor perhaps stand for long without obstructing the view of others. Its symbolic form perhaps supersedes its physical form: carved poles, unlike other monuments, intentionally carry a sense of mortality within them – they are made of wood, not stone. One does not sit at the base of a pole but, rather, one goes about one's business, one might say, within its shadow. All of these things affect how the pole is taken into the park space, and how it in turn changes it. When understood this way, this monument oversees daily a child's play area, an office, a kitchen, an activity room, several benches, cherry trees, grass, and any number of small gatherings of friends and strangers. It enhances a feeling of village in this already vibrant place. It is a well-recognized landmark within the neighbourhood, but is almost unknown outside it. It has truly succeeded as a monument done *by,* not *for,* the many peoples of the Downtown Eastside.

PART 2

FRAME

Marker of Change, the CRAB Park boulder and *Standing with Courage, Strength and Pride* were installed between July 1997 and June 1998, as shown at the locations on the map on the following page. However, these dates and the close proximity of the locations, while they highlight some of the commonalities of the monuments, also mask key differences. For example, consider the differences between the cadastral map (on p. 90) and the reproduction of a hand-drawn community map completed by residents in a community agency within the Downtown Eastside (on p. 91).

The second map reveals a nuanced understanding of the public spaces and places in the neighbourhood regarding notions of significance and belonging as well as levels of formality, safety and surveillance. The second map is thus more revealing of the worlds in which the monuments were conceived, contested and built, and in which they continue to be used.

Unpacking this more nuanced map requires that I temporarily set to one side the immediacy of the accounts of the three monuments presented in Part 1, so as to place them within a larger, more conceptual context. This context is informed by bringing a geographic sensibility to an intersection of several literatures. It is my contention that in order to understand how monuments work in social settings, it is important to first consider the elements that monuments seek to entwine: public space and social memory. Though both appear intuitively obvious, in that most of us encounter and perpetuate them every day of our lives, I want to inspect them each more deeply in order to lay the framework for how and why the Vancouver monuments represent such extraordinary challenges to conventional monumental norms. When, at the close of this section of

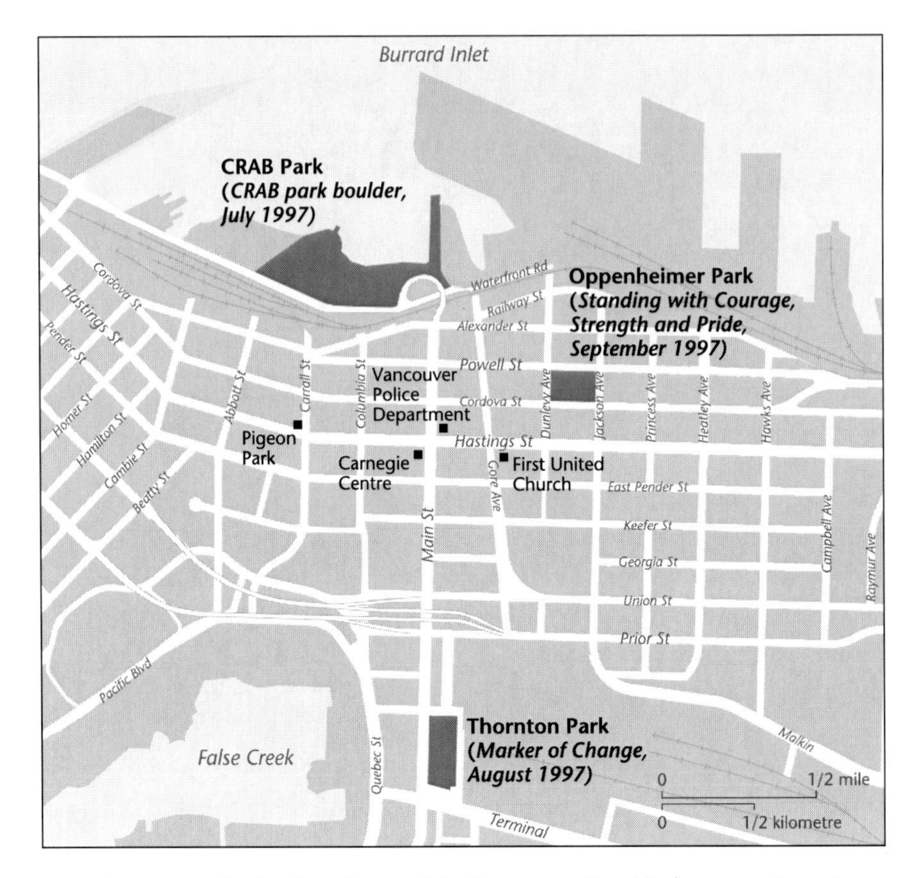

This is a standard cadastral map of the Downtown Eastside. | *Cartographer: Eric Leinberger*

the book, the Vancouver monuments re-enter the discussion, the details of their origins, installations and uses shift from a simply narrative focus to a more sharply defined analysis, indicating exemplars of counter-hegemonic, creative manipulation of the monument form and illuminating what I propose as a "politics of visibility."

Public Space, Social Order and Visibility

It is perhaps most useful to begin with the basics. Although understood intuitively by most lay persons, "public space" is a difficult concept to define. Vernacularly, legally and tacitly, "public" (space, sphere or realm) is conceived as one-half of a binary, where "private" (space, sphere or

This map was developed as part of a Community Directions group project, wherein participants were asked to place features on the cadastral map. The result shows a very different relationship between the neighbourhood's spaces. | *Source: Community Directions Community Map; Author's field notes*

realm) is the other. These (contested) categories define boundaries between households, market economies, the state and political participation. In other words, they delineate both spatially and metaphorically the different activity arenas that characterize a life of emotion, commerce, social reproduction and respite. However, these highly abstract categories are predicated on profound misreadings and exclusions of the realities of gender, class and racial aspects of social relations. For these reasons, feminist scholars in particular have questioned the entire practice of normalizing such dualisms, insisting that to either accurately describe or helpfully analyze social relations, a much more nuanced approach is required. What these scholars advocate instead is a research stance that discloses, details and acknowledges the researcher's particularistic point of view – as situated, contextualized and necessarily partial, rather than objective and omniscient, via the "god-trick of seeing everything from nowhere" (Haraway 1991, 189).

But that still leaves the problem of defining "public space." If it cannot be accurately defined in abstraction but can be defined only for a specific time and place, how can it be usefully employed as a topic for analysis? While in the classical world public space indicated a site of freedom and permanence in which political debate could be entered into and displayed (by some), in contemporary cities public space seems to exist much more problematically. Public space now exists along a continuum of coercion and exclusion mediated by legal restrictions, which includes bubble zones around abortion clinics; complex public/private ownership of parks; the mandated activity of consumption as in malls and historic districts; the development of analogous "public spaces" that are in reality more open than truly public; and the omniscient presence of surveillance technologies.

Still, our daily lives tell us that public spaces exist, and are full of bodies pursuing a variety of tasks, relationships and purposes. These public spaces seem to function as sites wherein civil society and civil inattention are cultivated. Such areas also serve as sites of display and/or theatre. This combination – of individuals, social interactions and visibility – plays a critical role in negotiations of identity, both of the individual and of larger social bodies. In spite of the understanding that "there are not and never have been any truly open public spaces where *all* may freely gather, free from exclusionary violence" (Mitchell 1996a, 130; original emphasis), the fact that our lives include public spaces suggests that it is less useful to try to define what public space actually *is* and more useful instead to observe how public spaces *function*.

Many of us are "moved on" from public spaces, by both overt means (bylaws, arrest and physical force) and more subtle means (harassment, threat and refused service) – especially the deranged, the infirm, the young, the elderly and the violent – but often, too, the female, the non-WASP, the non-heterosexual, the non-consumer. While such exclusions may be overtly marked or enforced, they are also internalized, as often noted by feminist scholars, as necessary to promote feelings of safety, or as adjustments made to conditions of urban voyeurism. But people have a more complex relationship with public space, one that extends beyond their personal experience of it. Public space is profoundly implicated in the process of "othering," that is, of defining not only what (and who) is dominant, but what (and who) is deviant, and holding these elements in a hierarchical tension centred on the illusion of agreement about social norms of non-disturbance. This view of public space

> implies a tacit agreement, a non-aggression pact, a contract, as it were, of non-violence. It imposes reciprocity and a communality of use. In the street, each individual is supposed not to attack those he meets; anyone who transgresses this law is deemed guilty of a criminal act. A space of this kind presupposes the existence of a "spatial economy" closely allied, though not identical, to a verbal economy. This economy valorizes certain relationships between people in particular places (shops, cafes, cinemas, etc.) and thus gives rise to connotative discourses concerning these places; these in turn generate "consensuses" or conventions according to which, for example, such and such a place is supposed to be trouble-free, a quiet area where people go peacefully to have a good time, and so forth. (Lefebvre 1991, 56)

When there are obviously Others present in these "public" spaces, as Ruddick (1996) points out, intersecting constructions around race, gender and sexuality can exponentially affect exclusionary dynamics. The addition of economic disparities furthers these exclusions and deeply affects the placement of people along power geometries (Massey 1994), social gradients that can profoundly influence degrees of mobility, access, choice and agency across a range of scales linking global and intimate realities. These tensions are played out in many ways, but along their most visible avenues are parks, streets and plazas. Thus, "public space," which sounds less restrictive than "private space," is ironically "very much a space of and for control ... For parents, for children, for men and women, public space therefore remains a highly ambiguous place, presumably a place of freedom, maturation, exploration, and so forth, but also a place of restriction, of all manner of unwritten rules and regulations governing interaction" (Mitchell 1996a, 130).

And yet, as Don Mitchell acknowledges, though truly public space does not really exist, "it is always there for the making." This touches on a few critical aspects of public space: the idea of establishing public space remains a fond hope; public space remains an ever-present potential site of agency; and even festooned with restrictions – "always and inescapably a product of social negotiation and contest" – public space remains, at least somewhat, a social reality (Mitchell 1996a, 130). These simultaneously held, seemingly contradictory ideas about public space – that it is a real entity but also a necessary illusion; an economic category nevertheless saturated with cultural symbols; a laundry list of what's left over after the private is negotiated but also an incubative resource for shaping a collective civil society – compel us to look more closely at how public space, as an idea and as a reality, is produced and reproduced in our daily activities.

Production of Space

In his analysis of the production of space, French philosopher Henri Lefebvre (1991) explores how capitalism's various stages have been spatially expressed. Beginning with the notion that capitalism necessarily involves economic violence, he also claims that capitalism extends that violence to inscriptions in space via a masculine "phallic brutality" overwriting social life and social space, understood here in both physical and metaphorical senses. He contends that specific modes of (capitalist) production have progressively "de-corporealized" social spaces, and that capitalism "depend[s] not only on a logic of accumulation but also on a *logic of visualization* through which human spatiality b[ears] less and less relation to the human body" (Johnston, Gregory and Smith 1994, 481-82; original emphasis). The result of this progression of increasingly de-corporealized space is the colonization of older, historically sedimented spaces via spatial practices that use both technologies (for example, cadastral mapping) and representations of space to install an abstract space – "space represented by elite social groups as homogenous, instrumental,

> "[In the social sciences] space was treated as the dead, the fixed, the undialectical, the immobile. Time, on the other hand, was richness, fecundity, life, dialectic."
>
> — Michel Foucault, *Power/Knowledge*

and ahistorical in order to facilitate the exercise of state power and the free flow of capital" (McCann 1999, 164).

This abstract space is the space associated with the conflation of hegemony, permanence and visibility (all, of course, aspects of conventional monuments). Seen as both transparent and inert, abstract space implies that space itself is merely a backdrop to more important, and more efficacious, social processes.

But if space is perceived merely as a backdrop, the act of perception itself must be scrutinized. Several scholars have written about how the "scopic regime" (Jay 1992) disciplines both what is perceived and the orientation of the viewer to the viewed. What is cast as passive, as framed, as foregrounded – even the very idea of perspective itself – has deep ontological implications. When is something properly composed? Where do, for example, monuments "belong"? What Lefebvre emphasizes is that preferences for visual ordering that are present in traditional monuments – that is, those that are predicated on distance and relations of mastery and verticality between the observed and the observer – reveal deeply unequal power relations that are powerfully inscribed and re-inscribed even during the apparently innocuous act of looking. Thus the norms of non-interactivity, non-emotion and an individual's subservience to a larger social order are all wordlessly reinforced.

For Lefebvre, to shatter and dislodge abstract space and the dominance of the logic of visualization (for him a political as well as a personal goal) is to reincorporate the body via counter-discourses of what he calls the spaces of representation. His notion of spatiality thus links mental space with material space, and reimagines a spatiality that celebrates the body in all its dimensional sensuality, including its historical memory of lived, spatialized experience. Lefebvre sees this reconnection of materiality in everyday spaces as providing an emancipatory possibility for social life with both *humanist* and *political* implications for public space. In other words, public spaces are a crucible for negotiating and performing the social self, both in the sense of individual identity and also in the sense of relational identities.

The challenge, then, is to define public space in a way that incorporates not only lived personal experience but also the cognitive and social experiences of how public space functions and how it is imagined to be. Further, a definition must allow for the recursive processes of producing and reproducing space. That is, if public space is *not* an ahistorical, homogenous, inert backdrop to social relations but, rather, an integrative, dynamic constituent *of* them, a definition needs to embrace public space conceptually far beyond either the empirical or the psychologically based experiences of it.

With this in mind, I propose that public space be understood as a dimensional, not a singular, entity. I would suggest, as a heuristic device, that public space exists in three domains. The first domain is imaginative – an internalized, pre-verbal, conceptual space, an ideological space in which what is possible is contemplated; it *may or may not be* transformed into commitment to the other two domains. Despite its abstract character, the first domain is nevertheless a domain of public space because it involves an awareness of "self-with-others" in various imagined communities (Anderson 1983; Hall 1995). Thus, when one contemplates an outdoor public marketplace, one imagines a public space. The second domain is discursive, designating the social, interactive realm of ideas shared and committed to language and imagery. This domain includes maps, lyrics, photographs, paintings, planning diagrams, texts like novels and images such as films and tourist brochures. This domain is where ideas of public space are fashioned into some means of representation, which can in turn be communicated or exchanged. The third domain is the tangible, physical realm, experienced empirically. In this domain public space is apprehended through all the body's senses. In cities this domain includes those locations that come to mind as randomly accessible and inhabited by a variety of users for a variety of purposes and encounters of "civil inattention." Perhaps this is best defined in Erving Goffman's original description, despite its gendered orientation: "One gives to another enough visual notice to demonstrate that one appreciates that the other is present (and that one admits openly to having seen him), while at the next moment withdrawing one's attention from him so as to express that he does not constitute a target of special curiosity or design" (Goffman 1963, 84).

Use of Public Space: Eliminating Possibilities

Public space is the site (in all three domains) where the personal and the collective intentionally mix. Though many scholars underscore how necessary the random rather than controlled mixing with others is to the health of civil society, sociologist Richard Sennett, in his early work on suburbs, identifies one explanation for how this mixing enriches not only the individual self but also the social body. In *The Uses of Disorder* (1970), Sennett argues that the initial impulse to control one's mixing with others arises from an imbalance between one's access to and one's ability to use resources. Sennett characterizes this impulse as intrinsic to the state of the adolescent. He suggests this leads to a peculiar kind of fear and a set of behaviours that result in individuals choosing avoidance of risk as a social strategy for dealing with difference. Avoiding difference, and avoiding mixing with others except on one's own preferred terms, is designed to establish "a fixed order of life so as to transcend experiences of pain, dislocation, [and] being overwhelmed" (16). While this approach may be understandable in adolescence, if the behaviour becomes fixed in adulthood, it weakens an individual's social repertoire. It encourages the formation of an emotionally impoverished sense of self, which in turn results in the creation of counterfeit communities made up of collections of emotionally and socially stunted people. That is, *actual participation* in community *is replaced by an imagined state of belonging based on eliminating what disturbs*. As such, the limits of imagination and risk become the basis on which one associates with others, rather than, say, exploration, curiosity or desire. When these associations of avoidance (that is, imagined public space) are reified in the material environment (physical public space), a paradoxical effect can result: the "signs and images of [social] contact [appear] as more natural and desirable than contact itself" (Mitchell 1995, 120). Mitchell, for example, cites malls, corporate plazas, festive marketplaces and theme parks as apparently open, public spaces that are actually highly regulated. Note again how this links the issues of individual identity with the development of civil society.

Sennett was writing in the 1960s, and, like urbanist Jane Jacobs, encouraged a celebration and enhancement of (material) urban public spaces precisely because their vitality would in turn make more resilient communities, societies and minds. Despite the celebration that greeted their work at the time, our modern cities reveal that their ideas didn't prevail. And so here we are nearly half a century later trying to conjure up other means to dismantle brittle, fearful discourses of counterfeit community ("freedom-loving people versus the axis of evil") and nurture new forums for mixing with others.

In an urban context, public space by any measure is in flux. In theory, access to the various domains of public space is more open than ever, with the availability and relative ease of travel, the increasingly important role of cyberspace, and the plethora of media, imagery and ideas circulating around the globe (Appadurai 1990). In practice, however, globalization has etched fragmented yet entrenched and uneven patterns that actually decrease our chances to openly interact with others in truly public spaces. All three domains of public space are under assault externally from the institutions of commodification, militarization and surveillance. And now, post-9/11, and after the bombings in London, Madrid, Bali, Dar Es Salaam and Mumbai, not to mention the daily violence in any number of official war zones in cities elsewhere, political leaders exhort us to "go about our lives" but also to cultivate an internalized, self-disciplining fear of public spaces, where anthrax spores or snipers or shoe bombers or worse are lurking.

Uses of Public Space: Emancipating Possibilities

There is some indication, however, that there are also emancipatory uses of public spaces, and that these uses figure in both the strategies of the powerful and the tactics of the marginal in some interesting inversions and reversals. When a reigning monarch such as Jordan's Queen Noor walks among her people to lead a street protest about US policy and human rights, something is afoot. The international non-profit organization

Greenpeace consciously seeks to retain a grass-roots image of activism with its street-based canvassers even though it has an operating budget and global organization that exceeds, in personnel, equipment and tactical expertise, the resources of many developing nations. What is intriguing is that these apparent inversions of the trappings of power are *designed* for public space in ways that transcend simple concerns for performance. Perhaps something is happening to the way in which public space is perceived as a political resource, as a means to enhance or diminish hegemonic relations. If it is true that hegemony operates not as a use of power *over* (visible rule, or coercion or physical force) but as a use of power via the tacit (manufactured) consent of those oppressed (Gramsci 1971), then public space is too vital and visible a resource to leave unexplored, a mere backdrop to social relations. Particular kinds of acts – especially those that are visible, disordered, dramatic and designed for public spaces – are critical to the negotiations of hegemonies and to the potential for counter-hegemonic messages to take hold.

Some geographic scholars have looked at precisely this relationship between public space and hegemonic order. For example, the origins of the free speech movement in the United States show that the rights now legally enshrined in the US Constitution were intrinsically connected with the *situated* dissent in public spaces by unionized and striking workers. Supreme Court Justice Brennan's realization that there were inequities in access to "the marketplace of ideas" (and, significantly, even to the physical location of the marketplace) made him reflect that law could not protect the right to speak in the absence of also protecting the (physical) opportunity for the right to listen, and the (physical) opportunity to be persuaded: "The dissemination of ideas can accomplish nothing ... if the otherwise willing addressees are not free to receive them. It would be a barren marketplace of ideas that had only sellers and no buyers" (Mitchell 1996b, 170).

Thus, access to public space is again seen as critical to negotiations of both the self and the self-among-others. Yet because public space is imagined as a space of inclusion and equality but has always fundamentally involved exclusions, "to make excluded voices heard ... requires drama" (Tushnet, in Mitchell 1996b, 157), and drama includes more than speech, and more than appearance. It includes challenging – embodied physically in the streets and discursively in the courts and the media – which behaviours are appropriate in public space and which are possible. It includes, therefore, the Lefebvrian ideals of re-corporealizing public spaces: it is not enough to visually apprehend a public space; a public space must be experienced by the sensual, remembering body.

Activists across the political spectrum have used public spaces for this kind of drama, where the visual image of the public space is permeated by the particularity, sensuality and lived experience of specific human bodies. The AIDS quilt, for instance, conceived initially for the public space of the Washington Mall (a visually compelling nationally symbolic public space in Washington, DC), is deliberately designed with the human experience in mind. Each "panel" of the quilt is represented by one body, and the quilt must be reassembled, every time it is displayed, with live bodies. In its fragmented touring (it has rarely been assembled in its entirety), it dramatically suggests the fractured whole, and the presence of the quiet, deadly embodied pervasiveness of AIDS (Sturken 1997). Other visual, embodied dramas are taken up across the political spectrum: women encircling the Pentagon, tea parties at Greenham Common, praying by fundamentalist Christians around abortion clinics, women standing as the waters rise around them at the Naramada Dam, the Women in Black in silent weekly vigils on Jerusalem's street corners, homosexual "kiss-ins" in German parks, the weekly parades of Las Madres de la Plaza in Argentina, monks immolating themselves in Vietnam, monks chanting through the streets of Rangoon, the Orange marches in Northern Ireland, the Billboard Liberation Front of San Francisco – all of these speak to the use of visibility combined with embodied drama as a way of animating public response on a scale greater than the individual but less than the state.

Such incursions into the three domains of public space are both bracing and provocative, although it is striking how episodic and temporary they are. In their immediacy and emphasis, they form extraordinary moments. Of course, this is entirely the power of ritual – to use transience powerfully. Lefebvre calls such moments "trials by space," but cautions that "ideas, representations or values which do not succeed in making their mark on space ... will lose all pith and become mere signs, resolve themselves into abstract descriptions, or mutate into fantasies" (Lefebvre 1991, 416-17). Such episodic moments are also inherently a concession, an acknowledgement that it is incredibly difficult to sustain a counter-hegemonic visibility – one that will not simply be absorbed, eviscerated and accommodated once it is drained of the potency of surprise (Stallybrass and White 1986). The issue is not, then, simply to *be* in public space – fleetingly – but rather to occupy it on one's own terms, to control one's passage through it and to be able to linger in it undisturbed, even relatively permanently. But there are paradoxes to such continuous visibility.

One of the most powerful ways hegemonic messages are tacitly reinforced is through the construction of commemorative environments, which, again, enshrine visible (and simplified) points of reference as proxies for complex lived histories. The following account of a site in Lexington, Kentucky, shows how hegemony uses visibility and also hints at how such strategies can be undermined:

> From the 1830s to the 1850s, the city's courthouse square, in the heart of the contemporary downtown financial area, was a major regional slave market. As with courthouse squares throughout the U.S., the place is currently the site for the construction and valorization of the official history of the city through the strategic placement of historical markers, statues, and monuments. Conspicuous by its absence, however, is any trace of the trade in human bodies that was central to the city's economy in the mid-nineteenth century. Despite the complaints of many residents of the city, the courthouse square's collection of war memorials, markers, and statues commemorating, among others, local Confederate commanders has not been joined by an emblem of the suffering and subjugation of countless Black slaves at the hands of the city's merchant elites, nor by a monument to the bravery of those involved in the underground railway which transported slaves across the Ohio River to the north. (McCann 1999, 170)

Thus, what is marked in a public space may not only modify previous histories, it may also completely eclipse them (for example, the city's connection to the slave trade, or the slaves' use of the underground railway to the North), particularly if the discursive and imaginative domains of public space are not considered. Further, such exclusions and erasures in the physical environment can become naturalized, especially when there is a linkage between the commemorative environment and the performance of memory rituals (a present-tense mixing of imagined, discursive and physical domains of public space). Such rituals imply permanence and continuity (qualities ascribed to hegemonies), yet the histories they commemorate may actually be non-existent except as invented traditions.

The advocates for the three monuments detailed in this volume did not think in quite such theoretical terms about placing issues of social justice in the public eye. But they did understand the importance of permanence, of fixity, of lingering in public space rather than moving through it ephemerally. They did understand that inserting tangible things into public space can provoke discussions and ideas, and they set about doing this through public education prior to the installation (as in the case of *Marker of Change*) or with inscriptions (as in the case of all three). The advocates did understand about interactivity and accessibility, as attested both by the designs of the monuments and by the rituals conducted at each. And they certainly understood the relationship of public visibility to social legitimacy (even, perversely, receiving death threats as they organized to protest against irrational violence in the community). They also understood that memory played a role in establishing that social legitimacy. And so it is useful to now turn to a close discussion of memory in this context.

Memory: Blending the Personal and the Social

Memory studies have proliferated in recent scholarship, particularly among historians and anthropologists. Just as with public space, however, there is an intriguing divergence between the lived experience of memory and its academic analysis. Most of us would consider memory an intrinsic part of daily life, affecting "everything from the ability to perform simple, everyday tasks to the recognition of the self. Memory establishes life's continuity; it gives meaning to the present, as each moment is constituted by the past" (Sturken 1997, 1).

Personal memory functions to organize competing sensibilities of time, place and circumstances. When we consider the etymology of "memory," we can see how various cultures have understood the relationship between memory and sensibility. In Old French, the word for "mindful" *(memor)* is akin to the Greek *memeros* ("anxious"), *merimna* ("thought, solicitude") or *martys* ("a witness"). Similarly, the Sanskrit *smarti* translates as "I remember." Other root words suggest more the dimension of time: Old Irish *maraim* ("I remain") or Latin *mora* ("delay"). All of these examples underscore the fact that personal memory links *relational* experience, either between people or between periods of time. Thus, in a personal sense, memory may serve an educational purpose (remembering the stove is hot when that button is red), it may set time horizons ("that was then, and this is now"), and it may serve as a repository of moral insights ("I will never make that mistake again"). In each memory function, the individual positions herself within a larger context – either of the self juxtaposed with a previous self, or of the self among others. In these ways personal memory connects an intimate pre-verbal world with a social world – the function of an imagined domain of public space.

One of the interesting things about memory is its relationship to reliable notions of accuracy or reality. Meghill (1998) argues, for example, that what matters most about memory may not be primarily the content or accuracy of something remembered but, rather, how memory functions to stabilize identity, by laying claim to authenticity and truth (in contrast with the historical account, which itself is always subjective and incomplete), or to serve beyond either tradition or nostalgia to allow fragments of the past to live on into the present. Meghill contends that the recent increased interest in memory studies may be "in part ... a response to the anxiety caused by the failure of modernity and of the scientific enterprise ... to provide an adequate account of that which, although past, continues to haunt the present because it represents things that have been pushed aside without being confronted and resolved" (56).

Another challenge to the efficacy of memory as a phenomenon is that it is exceedingly mercurial. Metaphors and mechanisms for depicting memory have long been imagined as locations, but locations with very particular characteristics: "Over the centuries, [the place where memory is stored] has been imagined as a waxen tablet, an electrical trace, a cluster sparking on a network, a library made of eroding fabric, a mental theater with painted doors. In practically every version, the site is 'built' from a highly malleable substance, or moves along a very slippery trail. Memories tend to efface easily, or lose track" (Klein 1997, 14).

Klein's description stresses not only that memories are elusive entities in themselves but also that they interact complexly with each other. The novelist Jorge Luis Borges recalls his father musing about memory:

> For example, if today I look back on this morning, then I get an image of what I saw this morning. But if tonight, I'm thinking back on this morning, then what I'm really recalling is not the first image, but the first image in memory. So that every time I recall something, I'm not recalling it really, I'm recalling the last time I recalled it, I'm recalling my last memory of it. So that really, I have no memories whatever, I have no images whatever, about my childhood, about my youth. (Borges, in Klein 1997, 318)

Problematically, it is not established whether or not even so-called source memories themselves are ever intact. In his seminal work on how individual and collective memories are constructed, Maurice Halbwachs (1952/1992) argues that it is primarily through (contemporary) membership in class, religious or national groups that people not only shape and interpret but actually *create* the content of both individual and social memories. He asserts that what influences the (memory) content "retrieved" is less based on any original source event than it is on contemporary social contexts, anxieties and alliances. Thus, what is remembered about the past is actually more about the present. But as Meghill cogently argues, Halbwachs' conceptions rest on another unsettled proposition: "for Halbwachs, the social identities in question already have a determinate existence before the collective memories that, at every moment, they construct" (44). So which comes first? Memories or identities?

As if all this conceptual slipperiness weren't enough, there is an additional complicating factor: memory ironically involves enormous amounts of forgetting. There are different explanations for why this is so. One is that humans would simply be overwhelmed if we could not forget far more than we recall; patterns are impossible to discern without elimination of excess information. Second, just as we imagine that the locations where memories reside are themselves corruptible and unstable, so too are the boundaries and mechanisms of memory infrastructure subject to decay, and remembered details are reassigned with a rather alarming fluidity. Third, quite apart from the rearrangements of our memories of source events, (present) context and emotion complicate how the (partial) past is (partially) thrust into the present. That is, context and emotion influence what is "retrieved" by our memories and how (Whittlesea 1997). And who knows? Perhaps memories themselves have an animation that we haven't yet perceived.

Yet despite our uncertainty about what exactly memory is and how it works, we all seem to rely on memory in our daily lives, to reify its existence and purpose and to value it as a means to stabilize personal identity. Surely it also functions to stabilize collective identities – as detailed in Benedict Anderson's discussion of imagined communities (1983). True, some scholars lambaste the very idea of collective memory, claiming that "consciousness and memory can only be realized by an individual who acts, is aware, and remembers. Just as a nation cannot eat or dance, neither can it speak or remember. Remembering is a mental act, and therefore it is absolutely and completely personal" (Funkenstein 1989, 6). Yet this stance somewhat recklessly jettisons an important conceptual tool for analyzing social entities. The rituals, images, texts, invented traditions and cultivated practices around notions of the past, which groups activate repetitively over extended periods, function to consolidate status, hierarchies and boundaries. Memory certainly changes (when, for example, it shifts from personal to social memory), but this is because when memory becomes social, it shifts away from the solely imaginative domain towards the discursive and potentially physical domain; of necessity, social memory is *shared* and thus becomes inherently socially prescriptive. For example, anyone who lives through a war retains intensely personal mementos, images and stories. These may or may not be passed on to intimates and revisited in personal rituals and acts. But acts of social memory about wars (such as cenotaph ceremonies) are emotional *and* intentionally cerebral; they are *designed* for audiences of larger social bodies (for example, "future generations"). Rather than serving solely for purposes of reminiscence, they are also intended to impart moral instruction.

I want to be clear on this relationship of moral instruction, social memory and visible (especially relatively permanent) material expression. As noted above, public visibility is a powerful force for negotiating and contesting hegemonic relations. When, as in the case of Lexington, Kentucky's courthouse square, the strategic, visible and permanent presentation of history does not acknowledge a central human experience of a place (slave-trading, in the case of the Lexington example), the physical end result, ironically, suggests not the incompleteness of the display but, rather, the insignificance of what does not appear. In a visually literate society, gradations of visibility can be deployed to imply degrees of legitimacy, significance, centrality and marginalization. Thus, if points of reference of social memory that uphold the characteristics of abstract space can be installed, it follows that the ensuing moral instruction is hegemonic. If, however, the points of reference are destabilized – by the living memories of observers, for example, or by alterations to the seamlessness of abstract space – then the nature of the moral instruction changes. When buried or obscured memory is intentionally made visible, it can enhance, enrich and strengthen individuals, communities, identities, social justice and sense of place (Hayden 1995). Clearly, this is in part what the Vancouver monument organizers were trying to do: wrestle material forms into the public eye and tell a story not usually told in public space.

Telling memory-stories also calls into question ideas of performance, and of using the elasticity of memory intentionally. The following section on monuments will detail how this can be done in material form, but the practice can be perhaps more easily grasped by observing a skilled storyteller at work. For example, in her account of Mrs. Sidney, a Yukon elder, anthropologist Julie Cruikshank recounts how Mrs. Sidney told the "same" story over several decades but altered it for different social contexts so that it was successfully offered as a text, a gift, a settlement and a commemoration. As Cruikshank observes, there is a crucial difference between "what a story *says* and then what it can *do* when engaged as a strategy of communication ... meanings do not inhere in a story but are created in the everyday situations in which they are told" (Cruikshank 1998, xv).

Crafting Memory

Collectively, these observations about memories – that they are slippery, that they link the personal with the social, function differently when they are shared as opposed to when they are simply contemplated and function differently in various contexts – suggest there is an artfulness involved in the foregrounding and elimination of elements in the process of crafting social memories. Which details should stay, and which should be muted?

Here scholars disagree: some argue that the past is indeed "fundamentally knowable," while others posit that there is a "zone of incomprehensibility" lying behind what we are able to know (see Meghill 1998, 52-54 for a summary). But generally scholars acknowledge, albeit for different reasons, that accounts of the past necessarily involve eliminations or suppressions of thoughts, memories and events. Here, Freud offers a useful insight. He proposes that crafting memory engages a cognitive manoeuvre that permits "memories [to] function[,] to hide, or screen out, more difficult memories the subject wants to keep at bay" (Sturken 1997, 22). Freud calls this "screen memory." When an image is projected onto the surface of a film screen, certain details pass through the screen while others are reflected back to us as images. Freud postulates that difficult memories function the same way: disturbing memories can be subsumed (forgotten) by the more placid surface memories.

Applying this metaphor for personal memory in the context of shared memory, Marita Sturken (1997) suggests that the process of social memory crafting involves two stages. When a momentous event occurs, in order to reconcile themselves to it (and to participate in creating a shared "screen"), people must engage in a *collective, simultaneous remembering and forgetting*. First, they must experience a collective (or apparently collective) re-enactment of the event, as if they were a part of it. Second, there must be a public narrative-making that focuses not so much on accuracy as on reassurance: "The value attached to narrativity in the representation of real events arises out of a desire to have real events display the coherence, integrity, fullness, and closure of an image of life that is and can only be imaginary. The notion that sequences of real events possess the formal attributes of the stories we tell about imaginary events could only have its origin in wishes, daydreams, reveries" (White, in Sturken 1997, 8).

In other words, in crafting social memories into narratives that create a sense of coherence and ontological security, there is an attempt to create meaning out of disparate elements, to suggest a culturally appropriate trajectory of events and to evoke a sense of resolution – as stories, performances and morality tales do. A satisfactory process of social memory crafting thus allows the personal self and the social self to realign. Further, these shared processes of negotiating social memories allow for an apparent social continuity and a hegemonic consensus to take shape. As George Orwell famously wrote in *1984*, "Who controls the past controls the future. Who controls the present controls the past" (Orwell 1949, 35).

Crucially, however, satisfactory social memory crafting is not always possible. This is particularly pertinent in the aftermath of trauma – as in the case of all three monuments. Following excessive trauma, processes of social memory are sometimes "wounded" (Chivallon 2001), in the sense that either the foundational past event is so disturbing that the present actually cannot be reconciled with it *or* that the remembering process itself is so significantly disturbed that it is impossible to assemble a redemptive or even coherent narrative (in both cases, the "screen" is damaged). Chivallon illustrates her argument with the example of slavery. In the first sense of wounded memory, she notes that the sheer size and extent of the diaspora of African slave descendants profoundly shattered, both geographically and psychically, the ability of individuals to fully understand the scope and nature of their dispossession. For the purposes of our study, this kind of powerful disorientation would appear to be especially relevant in the case of the CRAB Park boulder, where the monument was established before the full extent of the disappearances was known, and the social rules governing the uses of names proved dauntingly complicated. But it also applies in the case of *Standing with Courage, Strength and Pride*, as, in recognition of the systemic nature of the intersecting violences of poverty, racism and addiction, the inscription simply refers to those "who have died unnecessarily and to those who have survived" – how could all of the trauma, and everyone who suffered its violence, be named?

In the second sense of wounded memory, there is a cognitive disconnection between the way people wish to imagine themselves and the way they have actually lived in the world. In Chivallon's example of slavery, she observes that because European ideals of "Liberty, Equality, [and] Fraternity" so permeate how Europeans wish to imagine themselves, it is impossible for modern Europeans to reconcile the fact that achieving these ideals actually frequently required gross distortions of them – that is, the denial of these very things to other humans. Resonances of this type of wounded memory dynamic may have been at play in the powerful denunciations and threats surrounding the inscription of *Marker of Change*. Recall the resistance to the explicit use of "men" in "women murdered by men." Why did some imagine *all* men were included in that inscription?

In addition to Chivallon's types of "wounded memory," there is a third dynamic at play in manipulations of social memory, one that also pertains to the Vancouver monuments. As in the case of the Lexington public square, there can be an intentional distortion or even destruction of social memory visited directly upon sites, by first preparing the (discursive, imaginative) ground with the argument that certain social conditions *require* material alterations. For example, in her discussion of the intentional destruction of the Halifax community known as Africville, Jennifer Nelson (2002) recounts the weird discursive logic that preceded its obliteration: "If a section of the city can be linked to a misery and a suffering [that is] 'obvious' and incurable, destruction can be looked on as a sort of rescue" (173). Such dramatic reshaping of social memory (through the construction or destruction of built forms) is, of course, more easily visited on "housed" monuments (see Till 1999), since their physical contexts are so much more amenable to reconfiguration. But there are also uncomfortable dynamic links between aesthetic activities, gentrification and the displacement of residents (see Deutsche and Ryan 1984; Deutsche 1988 and Burk 2006 for a related discussion).

Memory is not always weak, of course, even in the aftermath of trauma; there are those who fight through their memories of traumatic experiences, in order to "re-align" the self and society, and to assert that "memory *is* prevention," as one of the *Marker of Change* activists put it. In these battles, different mechanisms are employed. Again, when we consider how these processes influence social memory, rather than exclusively personal memory, it is important to look at the sort of memory crafting that occurs in public. The narrative crafting of social memory that is undertaken for a face-to-face gathering, as in the case of Mrs. Sydney, the Yukon elder, highlights the malleability of narrative, the way it can (and properly needs to) fold to fit different social needs, audiences and circumstances. But when electronic, simulcast media becomes involved, the process of assigning a narrative gloss shifts, and becomes more rigidly and continuously applied. When traumas are witnessed via media in real time, such as in the case of the collapse of the World Trade Center, or the bombing of Baghdad, signalling the beginning of an international war, there is a need for a memory-crafting process that, first, *includes* viewers via a virtual re-enactment (and a constant CNN-style replaying of the traumatic event) so as to make the physically dispersed audiences a common imagined community. Second, this allows interpreters to fantasize aloud various narrative scripts to psychologically "smooth" the trauma until a preferred meaning of the event emerges (Sturken 1997). This meaning, once narratively settled, will then be inscribed and re-inscribed. Media transmission thus provides a kind of temporally circumscribed discursive public space.

Importantly, broadcast electronic media positions individuals as passive and relatively powerless entities within an imagined community. Such smoothing and scripting of memory must be understood carefully. While a narrative gloss may create a sense of ontological security, it also not only implicitly calibrates the grand meaning of the event but situates individuals within a larger imagined community, in quite specific power relations. When a culture encourages historical amnesia, then, mediated memory scripting is deeply influential in terms of both social and personal identity. The scripting can encourage but also destroy opportunities for reflection.

This is what happened in the American media scriptings of the 9/11 attacks on the World Trade Center, for example. The relentless coverage of the initial attacks can be understood, somewhat, as the re-enactment of the trauma in order to make all viewers members of a mass imagined community. The media coverage that followed these early "re-enactments" is quite disturbing, for other reasons. New York writer Joan Didion remarked on how, within a fortnight, the narrative scripting of events had taken hold with a vengeance, even in New York itself: "I found that what had happened was being processed, obscured, systematically leached of history and so of meaning, finally rendered less readable than it had seemed on the morning it happened. As if overnight, the irreconcilable event had been made manageable, reduced to the sentimental, to the protective talismans, totems, garlands of garlic, repeated pieties ... We now had 'the loved ones,' we had 'the families,' we had 'the heroes'" (Didion 2003).

Then, there was an entirely bizarre explosion of media coverage (even in Canada) six months after 9/11, as though a year were too long to wait for an anniversary. This, even as victims' families complained that in the rush to commemorate, there had been insufficient time to grieve. There was even a "placeholder" memorial of columns of light. It was as though the need to make use of 9/11's political value in larger debates (on national security, war and the complete abridgement of civil liberties for many) meant that the victims' deaths became, somehow, more public ideological property than private events, something for all to consume, wrapped in flags and fear. Returning to the ideas of Sennett, we can suggest that there was a strong desire to frame community, via the manipulation of memory, based on what and who could be excluded. In the opinion of one commentator, the American media engaged in an overabundance of commemoration because it was "socially easier" to engage in events this way rather than to cultivate a historical analysis of 9/11 (Gitlin 2002).

Memory-Making and the Body

These examples emphasize that the crafting of social memories is not an innocent social act, but one that carries with it significant consequences. Because such actions suggest arrangements of social distances, set interpretive horizons, evoke boundaries and disguise themselves as consensual, they mix the personal and social worlds in prescriptive ways. It is no accident, then, that most social memory scripting is done by the socially powerful, to inscribe, re-inscribe and – through repetition – naturalize certain kinds of interpretations and social hierarchies.

It would be a mistake, however, to assume that social memories are crafted only in the infinitely plastic worlds of texts and images, which can be endlessly manipulated, edited, airbrushed and otherwise modified. Social memories are also crafted materially, and address the body. As humanist geographers proclaim, when the body – with its desires, emotions, memories, sensations and actions – meets "space," it is possible to transform the abstraction of *space* into *place* – the particular, the lived and the unique.

This is the realm of ritual, and of embodied, empirical, collective action. There are innumerable examples of social memory-making at this scale, from roadside shrines to silent vigils, and including all the activities noted in the ongoing uses of the Downtown Eastside monuments. Memory-making at this scale, however, alters and actually complicates meaning-making. Without a media voice-over that has rehearsed "the" narrative again and again (via trying different versions of gestures, inflections and silences to arrive at an acceptably neutralized meaning) – each individual is left to make sense of others' actions (and engage in her own) in plain view, without the opportunity to try things out covertly first. Such direct, more embodied engagement with memory scripting alters the sense of agency. Whereas television actually works to impart a memory-sense that one has participated in an event without actually having done so (Mander 1978), the immediacy of memory scripting in a present time/place means that such passive engagement is simply not an option: participation becomes necessary.

Thus the crafting and performing of memory interacts with public space in complex ways. Both public space and social memory-making are involved – imaginatively, discursively and physically – in the negotiation of personal and social identities. Both memory-making and public space can be used to reinforce aspects of hegemonic relations. "Abstract" space can be presented as complete, even as it obscures its own lived histories and reinforces a kind of de-corporealized spatial experience. The construction and performance of memory narratives can silence and obscure. But equally, both public space and social memory-making *can* be used to challenge hegemonic relations. The seamlessness of abstract space can be disrupted; narratives are malleable. Through engagement, agency and embodiment, transformation of even the most arid of public spaces is possible. But the most challenging and perhaps most effective way of countering hegemonic use of public space is, as noted above, not simply to *be* in public space – fleetingly – but rather to occupy it, for whatever duration, on one's own terms. For that reason, it is important to look at one means of such transformation – the way social memory is crafted in the physical domain in the form of monuments.

Monuments: Permanence and Memory

There are many public sites of memory, as diverse as moments of silence, parades, museums, days of fasting, memorials and monuments. These last two are often conflated. Although some scholars distinguish between them on the basis that memorials are related to past deaths and tragedies, while monuments are essentially celebratory, in fact mourning and celebration take place at both kinds of sites (Young 1993). Because of this, and because monuments and memorials are relatively more permanent than other expressions of social memory, I use the term "monument" here to include both monument and memorial forms.

Although monuments have existed in many cultures, they began to re-emerge at the end of the Middle Ages as distinct from the sepulchral and public decorative sculptures of the Renaissance and Baroque periods (Michalski 1998) and became, in Europe, sometime in the 1870s, "an artistic, political and social domain in its own right" (Michalski 1998, 8). At this time, they became closely linked to national and imperial projects. These "modern" monuments, like ancient ones, have been designed according to certain norms. Traditionally, monuments embody forms that uphold and re-inscribe the regime of de-corporealized looking. Those who build heroic and/or historic monuments seek to centralize, specify and impose explicit social messages on public space. Monk (1992) claims that this desire for such masterful commentary is an impulse of control, and of marking not only what is important and noteworthy, but also what is cast as passive, empty or uninteresting. She points out that, much as a frame differentiates what matters from what is background, so too are the physical design features of traditional monuments meant to stand out against "unimportant" immediate surroundings. It is this notion that underpins the use of monuments as devices by the most socially powerful, in any society.

Traditionally, monuments also typically feature flawless, allegorical and idealized forms, and embody a kind of dropped-from-the-sky aesthetic distancing due to the use of relatively permanent materials, heroic proportions and contrast with surrounding elements. They are designed to be visible from a distance, often mounted high above passers-by; if accompanied by words, they feature inscriptions intended to mute any conflict or controversy by striking either a redemptive or, alternatively, a succinct, reportorial tone. In short, the typical monument crystallizes many of Lefebvre's ideas about a visual logic of domination inscribed on the physical landscape.

Within the Western contemporary city, it is easy to conjure up examples of such objects: the cannon in the park, the general on horseback, the cenotaph, the statues of aristocrats and rulers. Such forms draw visual power not only from their design elements but also from their sitings relative to other elements of the built environment within cultivated landscapes. For example, the *Statue of Liberty,* by virtue of its immense size and placement – related-but-apart from New York City's built environment – creates a powerful impression: "She gluts the eye with a sense of power, springing from the sensation of seeing the future ... When we look across from the foot of the statue towards the massed towers of steel and glass and stone [Manhattan] ... we are looking at a future that has happened. We experience Manhattan paired with Liberty, twinned by upward thrust, by man-made origin, by vastness of scale" (Warner 1996, 13).

The *Statue,* as well as all those monuments in the mind's eye over a lifetime of encountering similar shapes in various cities, suggests that monuments "work" in a number of ways. First, monuments may be seen as a sort of manifest pivot – a fixed point around which other things move – between the unseen and seen, and between past and present. Though this point can be understood in a cadastral sense, it can also signify a more lyrical positionality. For example, in discussions of the inuksuit (which are crafted forms of natural materials that "act in the capacity of a human" in Canada's far North), Inuit elders claim such forms are places "where the spirits can come to feel the warmth of the living" when they need it (Hallendy 2000). Others provide a different insight into how monuments may fit between the secular and spiritual worlds. Historian James Loewen (1999, 37) notes that Kiswahili speakers from east and central Africa divide the deceased into two categories: *sasha* (the living-dead) and *zamani* (the dead). These descriptors acknowledge the tension between memories of those who have died recently and thus still "live on" in the minds of their contemporaries *(sasha)* and memories of more generalized ancestors, who are not so much personally remembered as revered *(zamani)*. Markers and events commemorating those in the *sasha* realm may be saturated with fresh grief and also controversy; *zamani* monuments, however, are not primarily motivated by loss or grief, and "usually go up to serve the political exigencies of the time of their erection" (Loewen 1999, 38). This observation seems to align with Western experience of monuments as well, and to suggest a monumental counterpart to Halbwachs' (1952/1992) initial thesis about how contemporary contexts affect individuals' memory. Often, the establishment of markers about social upheavals (the Scottish clearances, the World Wars, the Japanese internments in British Columbia) occurs not during the events they mark, or even immediately after them, but, rather, in the next interval of social spasm, moral panic or political effort. Thus, certain renderings of the past shore up contemporary alignments: "all monuments tell two stories. One about the event they commemorate, and one about when they went up."[1]

The creators of monuments thus seek to not only commemorate the dead but also educate the living. But crucially, this latter function is entirely dependent on one of two things: either that enough social context is shared by the creators and passersby that the monument's intentions are legible or that the previous knowledge and/or intellectual inquisitiveness of passersby is aroused sufficiently to engage in inquiry (Harjes 2005, 143). This desire for monuments to serve as a transparent instantiation of particular readings of events suggests that monument advocates intend to impress particular memories as a tacit or doxic reading. That is, as Pierre Bourdieu noted (1977), when "what is taken for granted in the natural and social world appears as self-evident" (164) and "within the

1 James Loewen, personal communication, May 30, 1999.

limits of the thinkable and the sayable" (167). One historian remarks that "to commemorate is to seek historical closure, to draw together the various strands of meaning in an historical event or personage and condense its significance" (Savage, in Younge 2002, 13). But as indicated in one extensive study of US historical markers, such renderings of historical events are often highly inaccurate. Remarkably, inaccuracy is often the point: Ernest Renan remarked to French nationalists in 1882, "Getting its history wrong is crucial for the creation of a nation" (Renan, in Bhabha 1990, 30). If monumentalism is used to impart a sense of authority and incontestability, it is thus less important that it be accurate than that it be resolute: "Public monuments are the most conservative of commemorative forms precisely because they are meant to last, unchanged, forever. While other things come and go, are lost and forgotten, the monument is supposed to remain a fixed point, stabilizing both the physical and the cognitive landscape. Monuments attempt to mold a landscape ... to conserve what is worth remembering and discard the rest" (Savage, in Osborne 2001, 14).

There is something, too, about the formality of monuments, and their overt permanence, that goes beyond commentary or expression and verges on moral instruction. The instruction intentionally uses aesthetics to impart moral qualities: "Historical monuments and civic spaces as didactic artifacts [are] treated with curatorial reverence. They [are] visualized best if seen as isolated ornaments; jewels of the city to be placed in scenographic arrangements and iconographically composed to civilize and elevate the aesthetic tastes and morals of an aspiring urban elite. This [is] an architecture of ceremonial power whose monuments [speak] of exemplary deeds, national unity, and industrial glory" (Boyer 1994, 33-34).

This (apparent) imperturbability suggests that monuments themselves may be used strategically in the material as well as the discursive and imaginative domains. Perhaps the clearest example of this is David Harvey's (1979) essay on the origins of Paris's Basilica of Sacré Coeur. On the surface, this edifice is an exquisitely beautiful and ornately detailed religious site, apparently created in a gesture of passionate religious devotion. But an "excavation" of its origins reveals that it blended the military, religious and state symbols of the day in a compelling and nearly seamlessly integrated display of unity. It also quite intentionally appropriated the site where the Paris Communards were martyred. Its placement thus *dis*placed alternate readings (and potentially incendiary memory rituals) at the site.

So not only do monuments occupy sites physically, but they can also function imaginatively as a kind of precipitate or fixer: as Holocaust memorial scholar James Young writes, monuments can "bring events into some cognitive order" (Young 1993, 5). This is particularly true for the more recent age of monuments in the North Atlantic world. The reemergence of monuments as a commemorative form coincided with a growth of nationalisms, when conflicts were depicted as being in service to larger sacrifices, and industrial progress was seen as a good in itself. At a time when new art forms responded to the acceleration of social life in the early twentieth century (consider revolutions in art, music, architecture and literature), monuments may have been employed – because of their fixity – in part to remind people of the idea of society itself, with its continuity and achievements. Certainly in recent years in the Ukraine, though statues of Lenin and Stalin have disappeared or have undergone extensive revamping (Komar and Melamid 1994), a variety of new monuments have gone up to celebrate the heroes particular to the region and culture of pre-Soviet Ukraine.[2] This suggests that a social need for monumentalism (that is, permanence, visibility, social memory) continues, though the preferred content may change.

Dotting the landscape with reminders of heroic deeds, events and people is a way of shoring up a kind of mythic landscape. In this sense, monuments can serve as nodes within larger framings (see, for example, Dwyer 2000; Heffernan and Medlicott 2000; Judt 1998), or, as Eric Hobsbawm (1995) puts it, offer up "an open-air museum of national history as seen through great men" (13). The prevalence and visual rhythm of a whole series of such physical, imaginative and discursive landmarks (with particular concentrations at edges and borders of territories) silently but powerfully underscores emotions of belonging and otherness, providing the individual with a sense of being part of an imagined collective.

2 Sean Markey, personal communication, June 29, 2002.

The way one imagines one's individual self in front of these more traditional monuments is as part, but a *subservient* part, of larger phenomena. Monuments are traditionally designed for spectacle; their very scale and mass ensures a theatre or stage-like quality to their presence. Commenting on the Third Reich's preferred style of monuments, one scholar observed: "[The monument] must be rigorous, of spare, clear, indeed classical form. It must be simple. It must have the quality of 'touching the heavens.' It must transcend everyday utilitarian considerations. It must be generous in its construction, built for the ages according to the best principles of the trade. In practical terms, *it must have no purpose but [to] be the vehicle of an idea.* It must have an element of the unapproachable in it that fills people with admiration and awe" (Tamms, in Osborne 2001, 12; my emphasis).

Because they are publicly accessible as sites of public expression, monuments are almost inevitably used as locations to signal contested meaning-making. It is the very overtness of their statements – either by placement or by form – that invites the inversions of their (intended) mastery.

This aspect of monuments reminds us that monuments do function in some ways as instigators of slow conversations. Consider in comparison a parallel art form: urban graffiti. While graffiti is intentionally a much more ephemeral (and thus socially plastic) form than monuments, both graffiti and monuments can serve as locations in a visual dialogue of "speech acts" and "language" (Cresswell 1998; Giddens 1977). Though "all monuments are efforts, in their own way, to stop time" (Levinson 1998, 7), stopping time is, of course, entirely impossible. What *is* possible is to stretch the act of speaking so that face-to-face encounters are no longer necessary – indeed, are neither intended nor desired. Unlike most art, (classical) monuments are rather odd creations, intentionally designed to mute attention to their own materiality and so somehow suggest self-evident and non-problematic acceptance of regard for their subject matter. Further, they are designed to "speak" to strangers in an entirely unobserved way. This inevitably makes for strange conversational rules. Conventionally understood, conversation is something like a ping-pong game: one comment is launched and another is sent back in answer. Gradually the dialoguing players move in flexible responses around the "court" of the entire exchange. With monuments, conversation is better understood as an encounter with a ball-throwing machine that launches ball after ball from exactly the same location; the impetus is on only one player to alter the interaction. In this way, monuments invite re-inscriptions, otherwise the dialogue falls flat.

This returns the discussion back to the original idea of the pivot, but with a slight shift. Rather than functioning simply as locations, I would argue that monuments serve also as, potentially, locations charged with the significance of signalling negotiations between hegemonic and counter-hegemonic forces. The signalling that goes on at these monuments may be highly scripted (as it is at cenotaph ceremonies throughout North Atlantic countries on November 11), but it may also be amateurish or incomplete. Or it may offer an intentional, parallel disruption. For example, in a repeatedly economically brutalized region of England, on a scrap of ex-industrial land, the incongruously massive *Angel of the North* rises, the largest structure between Durham Cathedral and the Tyne Bridge. The design evokes the keen religious significance of Christianity to this area, and also incorporates the craft knowledge of the steel and shipbuilding history of the region (Campbell, in Gormley 1998, 91). Elsewhere, the *Another View Walking Trail* in Melbourne, Australia, places markers of Aboriginal heritage juxtaposed literally alongside colonial markers (Jacobs 1998). In cases where social memory has been particularly disrupted or wounded, the boundary marking may be especially nuanced. As I observed at Cape Town's extraordinary District Six museum, residents of the formerly integrated District Six neighbourhood who were forcibly removed by the apartheid regime to various townships according to 18 invented racial categories, have been invited, with the advent of the ANC government, to come back. They revisit detailed replicas of their streets and homes, and add their narratives to the histories of exact places and times, directly alongside the streets they once inhabited, and which are now razed, the land discarded. Similarly, Chivallon's work on wounded memory is based on her observations of the building of the *Slave Trade Trail* in Bristol. She argues that the trail was intentionally designed not so much to change the city form itself but, rather, to change "the framework for reading what is seen" (Chivallon 2001, 354). She notes, however, that challenging the two versions of wounded memory (of, respectively, the Caribbean and the European peoples) needs time, and a variety of tactics:

> It seems that the work of restoring a difficult, shameful past that must be extracted from a dense silence could not be done unless physical traces in the urban fabric were made visible. *It was as though the past could not break through if restricted to the realm of talk alone; the very form of the city had to have its say.* This was so not simply because urban form has the power of the "visibility effect," but because form also brings about a necessary distancing between a present where harmony is wanted and an extremely turbulent past. Paradoxically, urban form – the symbols it displays and the relics it deploys – was charged with returning an unwanted history to collective memory, while at the same time keeping that very history at a distance, by *making it exist in stone and in monuments before making it exist in the community of "flesh."* (Chivallon 2001, 354; my emphasis)

These examples point to the difficulties of wrestling with what the narratives and expressions of social memory should be and suggest that even when a consensus is reached, there are terrific challenges to their physical design. Largely, this stems directly from the desire to impart a completion and a sense of authority. Monuments whose design suggests this kind of closure attempt to inscribe an apparently unproblematic set of norms: "This is what happened, this is where it happened, this is when it happened. And that's the end of the story." Yet there is a curious irony to such implied subtexts. Often, it is the repetition of rituals around a tradition that is actually regarded as more meaningful than the content of the rituals; in other words, the performance of meaning supplants meaning itself. In a sense, such "hollowness" extends also to monuments: "Great memorials are curiously non-committal ... the monumental monument tends to be, in this way, an open emblem. It tends to be FOR RENT" (Gass, in Warner 1996, 12; original emphasis). Monuments remain open to multiple interpretations; social memory is indeterminate.

Thus, to return to an earlier argument, not only is memory unstable – in storage, retrieval and purpose – but even the monuments themselves are unstable. First, they vary in terms of their capacity to project the desired screen memory narrative gloss if memory scripting is weak or accelerated. Second, they are not always able to transcend design challenges that stem not only from social relations but also often from prosaic logistics. There is a third instability as well: What happens when the apparent transparency and authority of a monument is threatened because its content turns out to be simply wrong?

Errors in monuments may exist within the design, within the manifest story or within the implications of the rituals attending them. Sometimes errors exist in all three. A classic example of this is New Orleans' 1891 *Liberty Monument* (Levinson 1998, 45-52), which commemorates an 1874 "victory" whereby the aptly named White League rose up and violently overthrew the government in power (a mixed-race alliance). Partisans of the White League insisted on the installation of an obelisk at the foot of Canal Street (a major throughway of the city), memorializing the event. In the coup, 32 people had died and approximately 60 more were injured, but the memorial was intended to honour only the 11 (Caucasian) White League victims. After the installation of the monument, an annual parade was established to reinforce this lopsided commemoration of the dead. In 1934, two large plaques were mounted on the monument, celebrating the general tenets of "white supremacy." These plaques were contradicted 40 years later by a third plaque, which stated that "the sentiments expressed are contrary to the philosophy and beliefs of present-day New Orleans." In 1981, New Orlean's first black mayor tried to have the monument moved, but the city council rejected his proposal. It did, however, allow "offensive wording to be removed." And so, up went a granite slab to obscure the 1934 plaques (rendering the wording of the later, contradictory plaque a bit confusing). A subsequent mayor, seeing the opportunity provided by a waterfront reconstruction project, tried again to remove the monument, but was blocked by a triad of traditionalists, historical preservationists and white supremacists. He succeeded only in getting the monument moved one block over, to a slightly less noticeable area, with the addition of one last plaque, which at least finally recognized, by name, 11 police officers who had died in the violence.

While errors in monuments may be unavoidable, research indicates that the extent of such errors is more than incidental. For example, a survey of several thousand US historical markers found that only a fraction were about events that actually happened, or represented with fidelity those persons actually involved. Far more common were creations of historical myth, changes to the gender or race of key figures, and even reversals of fact (for example, those who actually lost a battle were portrayed as victors) (Loewen 1999).

Such inversions and distortions may even accompany whole landscapes of preservation. The account of the demolition and reconstruction of the Texas *Alamo* is instructive: the alleged mass martyrdom of Anglo-Americans at the *Alamo* is regarded by the Texan tourist industry as "the birth of Texas" and "the definitive first step towards Texan independence from Mexico" (de Oliver 1996, 1). Central to the (re)telling of this story has been the reification of a moral and cultural difference between Anglos and Latinos – not coincidentally the two major populations in contemporary San Antonio. When the Daughters of the Republic of Texas (DRT) took over restoration of the site of the *Alamo,* they set about remaking the physical site to conform to the "perceived spiritual connotations of the Anglo cause," without any regard to historical accuracy: "The Anglo reconstruction plan required the radical reduction and demolition of much of the mission complex so as to make the ornate chapel façade the site's focal point. The fact that the chapel façade was immaterial to the siege of 1836 was irrelevant to the Anglo faction of the D.R.T. The chapel façade was emphasized as the symbol of the Alamo" (de Oliver 1996, 3).

Because very little commercial activity is allowed at the *Alamo* "shrine," de Oliver notes that the image of the *Alamo* chapel façade, endlessly repeated on the tourist products, is marketed as the (visual) anchor of the leisure and shopping environment (Riverwalk/Rivercenter Mall) which lies contiguous to the reconstructed *Alamo*. Thus, "the once traditional pilgrimage to a cultural site for spiritual or religious assets (the Daughters of the Republic of Texas' Alamo) ... has been replaced with a consumer pilgrimage (consumerism's Alamo)." Indeed, what has in many ways eclipsed any commemorative or educational value of such sites is that the performance of civil participation is conflated with simple consumption. This consumption itself can be a re-enactment of sorts, a simulacrum of participation (see Boyer 1992).

One of the interesting points in the history of the New Orleans' *Liberty Monument* and the *Alamo* is how the monuments' errors and the redrafting of the official narratives were shaped through physical restructuring or geographic relocations. As noted above with the cases of the Lexington Square, the Basilica of Sacré Coeur, and the reframing of Bristol, District Six, and Africville, *placement matters*. Placement can underscore, or undermine, social understandings. This suggests that we must investigate a geographic sensibility before returning to a fuller discussion of the Vancouver monuments.

A Geographic Sensibility

Geography is attuned to the nuances of placement, visibility and decay: there is a focus on the relations of centrality, periphery, absence, displacement, contiguity, proximity, erosions and frictions of distance over spatially variable topographies. Monuments serve as points around which these nuances are enacted socially in three ways: in the negotiations about what monuments should be, in the debate about where they should be, and in the performance of rituals around their planning, installation, reconsecrating and alteration. These negotiations and rituals all provide opportunities for memory-making and contestations of power and hegemony – performances, if you will, of social memory. Thus, an analysis that posits that public space is negotiated imaginatively, discursively *and* physically can highlight how monuments, though they don't determine social memory, do allow for people to create, recreate and signal versions of it. Also, though they may appear so, monuments are not socially inert. As shown by the AIDS quilt, the post-USSR placement of Lenin statues, and the recent destruction of the Buddhist statues by the Taliban in Afghanistan, physically permanent art animates social responses long past the moment of its conception or creation. Monuments' apparently inert forms may be ignored, but equally possible, their massive physicality may invite people to interrogate and problematize social meanings. Permanence creates shared spaces for memory-making and thus contributes to the belief in common memories – which some would say are a prerequisite for social order. By what they obscure and what they make manifest, monuments can reflect what is socially possible, what is warned against and what is dreamt.

Recently, several artists have sought to subvert the spatiality of the monument form. Consider the norms of traditional monuments: the high contrast with its surroundings, the fixity and heroic proportions, the bland inscriptions, the aesthetic distancing and, unlike in other forms of contemporary art, the propensity for monuments to subsume the contemplation of their own materiality. What if monuments were designed to directly invert these norms, to play with the monument form – suggesting its incompleteness, inherent errors, contradictions and decay, and juxtaposing it in unexpected ways?

Artists have been working to alter the formality of monumental norms for just over half a century. (Except where noted otherwise, the following examples are from Michalski 1998.) In 1951, deviating from the norms of figurative forms, Eduard Ludwig employed the first use of abstract forms in *Monument to Victims of the Berlin Airlift*. This was followed soon after by Reg Butler's artistically and substantively controversial *Monument to an Unknown Political Prisoner*, noteworthy because it was the first piece of public sculpture in the West intentionally created with two unorthodox viewing points – one from a distance (the monument was atop a hill, visible from both East and West Berlin), and one from directly beneath the monument itself.

A decade later (1968), Ed Kienholz created a devastating caricature of the iconic image of the flag-raising at Iwo Jima in his *Portable War Memorial*. Though this was a museum piece, its flagrant reference to war memorials arguably makes it part of this discussion; in viewing its inversions of many of the clichés of war memorials, "the mere idea of a patriotic war monument espousing heroic action becomes more and more ludicrous" (Michalski 1998, 176). Kienholz created a tableau that was intended to be "read" from left to right; in this way he overtly calls attention to the role of the observer in relation to the object. In the 1970s, other artists worked with this idea and took monuments in new directions.

One of these directions was invisibility. Though for almost two and a half millennia visibility was "considered as an essential prerequisite for public monuments," some artists wished to play with even that aspect of the form. Thus there were proposals for figurative monuments: in 1965, Claes Oldenburg's buried (upside-down) likeness of John F. Kennedy; in 1981, artist Timm Ulrichs similarly buried a likeness of himself. By 1988 this interest in invisibility had extended to entire structures, initiating a movement towards "negative" monuments. That year, Horst Hoheisel used the site of an old fountain, which was built by a Jewish merchant in 1908 and then destroyed by the Nazis in 1939, to bury a replica of the original. Speaking about the *Aschrott Fountain Monument,* Hoheisel said, "I have designed the new fountain as a mirror image of the old one, sunk beneath the old place in order to rescue the history of this place as a wound and as an open question" (Michalski 1998, 178).

Soon such direct appeals to the interactivity of forms and their observers led to conflation of (some) public installation art and monuments, particularly those pieces that were intentionally designed to unsettle social relations rather than provide closure. Thinking about memory and the function of monuments, several artists noted that conventional monuments often allowed the work of memory to be displaced to *sites* and *things* in a sort of catharsis. Such displacement removes the "response-ability" of memory from persons, allowing the memorial form (monument, event, text and so forth) to bear the work of memory. In this way "the scar does the work of the wound" (Wiseltier, in Novick 2000, 281). Attempting to counter this, Jochen Gerz and Esther Shalev-Gerz conceived of the Harburg Monument. Insisting that what they "did not want was an enormous pedestal with something on it presuming to tell people what they ought to think," the artists designed a steel pillar with a surface of soft lead on which passersby could inscribe whatever messages of contemporary reflections or memories they desired. In seven ritualized ceremonies over seven years, the scribbled pillar was sunk into the ground, leaving the process and the memory-sense of its presence, rather than its physical presence, as its monumental form. In other words, the memory work had to be borne internally, if it were borne at all. In another piece, Gerz, with the help of a class of students, removed over 2,000 cobblestones from a central plaza in Saarbrücken, temporarily replacing them with other stones for the few weeks it took to inscribe the original stones with the names of Jewish cemeteries. These original stones were then returned to their original (unmarked) locations in the square, amid much fanfare about the unveiling of a new monument. This too greatly complicated the place of memory: "Visitors ... may or may not realize they are walking on symbols of destroyed cemeteries. They may experience a certain uneasiness and may wish to avoid direct physical contact. This is impossible, however, as some of the stones beneath their feet have no inscription at all. There is no way to know the difference, no way to avoid contact" (Michalski 1998, 180-81).

Other examples of art that work to startle rather than aestheticize include the German admonishing *Mahnmale* memorial monuments. One of these monuments is a low stone form, apparently a bench in a public square, which only on close inspection is revealed to be a depiction of a Jew washing cobblestones (Czaplicka 1997). Similarly, the English socialist Sylvia Pankhurst erected an ironic anti-war monument outside her home, dedicated in 1935 "to those who upheld the right to use bombing planes."[3] Another is Hoheisel's *Warm Memorial* – a slab of concrete in the ground that is heated to the temperature of a warm human body. Passersby are literally brought to their knees in order to touch it. There are other possible subversions of the apparent transparency of the monumental form based on a de-centring of the acts of memory. Artists Renata Stih and Frieder Schmock installed simple benches, painted red, scattered around Berlin, with small plaques informing passersby that a particular act in the immediate area is forbidden to Jews. Thus, only as one walks around during the course of a day, searching for a place to sit, does one become aware of the message of the benches. Only cumulatively do their small, slight prohibitions impress on us how severely restricted and criminalized everyday acts – from the planting of gardens to movement in the streets – were for Jews under Nazi rule: eventually, the crime for Jews was simply to exist. These benches are in the tradition of the *Stolpersteine* ("Stumbling Stones") monuments, an art form building on Gerz's Saarbrücken plaza piece, using "generally small, rather nondescript pavement stones or street signs with an inscription referring to a past event ... often spread out over a city's streets and building façades ... placed at unexpected locations ... [and whose designers feel no] formal obligation to support governmental efforts to promote a particular vision of national identity" (Harjes 2005, 143-44). Since 1995, artist Gunter Demnig has installed more than 3,000 of these *Stolpersteine* monuments (Michalski 1998, 145). Another German artist uses a kind of monument form to

3 Richard Ross, personal communication, January 8, 2000.

reclaim the memories of the Jews murdered in the Holocaust by superimposing on shop fronts and apartments, on train platforms and sidewalks, projections of Jewish families at home, at rest, at work, in the very locations from which they were expelled (Attie, in Young 2000). Like Krzystof Wodiczko's *Homeless Projection* imagery on New York statuary (Deutsche 1986), Shimon Attie thus borrows the physicality of sites by adding images. A similar borrowed physicality occurs with the AIDS quilt, which, though made and remade in literally millions of locations, must always be reassembled to be viewed (Sturken 1997).

These examples suggest another movement in the artistic response to and use of monumentality. Called "counter monuments" (Young 2000), these expressions "[include] attempts to visually complement or change the appearance of earlier monuments" by adding details that contradict the original monument (Michalski 1998, 205). *These are often geographic interventions:* "The principal aim of a counter monument is to register protest or disagreement with an untenable prime object and to set a process of reflection in motion" (Michalski 1998, 207). Accomplished through a variety of strategies, including mirroring, transformation (as in adding figures to an original monument), immateriality and placement, all counter monuments eschew a triumphalist context.

Even state-sanctioned monuments *can* be counter monuments, and can redirect memory-making away from the monument and back to the social world. In the US, the most visited monument in Washington, DC, by far is the Vietnam Veterans Memorial, a wall that lists, by date of event, the names of all US personnel who died or went missing in Vietnam (Sturken 1997). Also referred to simply as the Wall, its design – that of a highly reflective, incised black V-shape cut into the earth of the National Mall – was chosen unanimously by the commission of Vietnam veterans, but has also been reviled for being about shame rather than glory, for disgracing the national stage (its two walls point unmistakably towards the Washington Monument and the Lincoln Memorial), and for commemorating, rather than healing, the ambivalence and divisiveness of the era. Detractors have, in fact, mounted two other (figurative, heroic, redemptive-type) monuments nearby to "correct" the Wall. During several field observations, however, I found that these two monuments were relatively disregarded by visitors. In contrast, visitors come in the tens of thousands to the Wall every day, week after week, year after year. There is a noticeable contrast between behavioural responses to these traditional monuments (snapshots, admiring glances, the reading of the plaques) and to the Wall, which requires that viewers enter the monument along a gently sloping path gashed into the earth. With each step, the visitor becomes more physically overwhelmed, the reflections of his/her own living body alongside those of other visitors reflected in the tall, polished black slabs of stone inscribed with the names of literally thousands of dead. At the apex of the monument, the walls loom far above the observer. There is an immediacy and vulnerability that astonishes; often people spontaneously begin to weep, tell stories, embrace strangers, kneel, pause, and touch the wall. People may leave notes, do rubbings or climb up the ladders provided to touch the names of loved ones. Many people bring their children here, and I have heard some visitors say quite extraordinary things about the Wall of Washington, DC: "Well, honey, because this was a war we probably shouldn't have been in."

These new kinds of monuments show us that though they may function to reinforce hegemonic arrangements overtly or covertly, or even accidentally, monuments can also call into question and challenge their own materiality, thus subverting the processes of memory-making and jarring the viewer out of a passive encounter with the monument into a more interactive one.

Permanence, placement and the controlled use of visibility are usually in the repertoire of the politically powerful. The powerful have, after all, the most to gain from upholding hegemonic arrangements of social hierarchies, uses of space, movement of citizens (and others) and rules of governance. But when a *partial submission* to the norms of monuments (for example, their fixity, their materials, their visibility) is used, along with *a simultaneous rejection* of other norms (for example, design features, inscriptions, placement), there are some interesting results. Monuments – which traditionally function to fix interpretations, arrest ongoing examinations of social memories, reify power relations, freeze notions of time and morally instruct passersby – can function instead to disrupt closure and destabilize doxic understandings of events, certain versions of history and ways of behaving in public space. The logic of visualization and its presentation of seamless abstract space, with its dynamics of

Vietnam Veterans Memorial Wall | It is remarkable how often visitors respond physically rather than just visually to this monument. | *Photo: Adrienne Burk*

mastery and subservience, can be breached. Public space – in all three domains – can be "disturbed." That is, attention can be drawn to what has been previously silenced, forgotten, eclipsed or skewed, through employing any combination of imaginative, material or discursive means. At the site of monuments, the random mixing of strangers increases. It is possible here to pierce the psychological processes that have ordered both what is remembered and how those memory narratives are crafted. The resulting insight into the machinery and art of such ordering encourages one to question how alliances and omissions of social connectedness are fashioned.

With these larger conceptual frameworks in mind, Part 3 opens with a closer contemplation of the implications of the community map that opened Part 2, and the ways *Marker of Change*, the CRAB Park boulder and *Standing with Courage, Strength and Pride* fit within this broader discussion of public space, social memory and a geographic sensibility about monuments. It is important to note that even the non-traditional monuments discussed in this section were commissioned, enjoyed state funding, sponsorship, and an imprimatur of legitimacy and/or have been feted within academic and artistic circles. If anything, the successes of these monuments highlight by contrast the remarkable accomplishments of the advocates for the Vancouver monuments, given their inexperience and obscurity, and the very real contemporaneous, unfolding drama that accompanied the monuments' creation and installation. There are lessons in the achievements of the advocates for anyone committed to greater social justice, more genuinely diverse communities and more robust public memories. I call this a politics of visibility.

PART 3
FORGE

In this section, the more theoretical ideas of the domains of public space, the malleability of social memory and the forms and norms of monuments are focused on the narratives of the three Vancouver monuments. In that these monuments were so unlikely – in their conception, advocacy, intentions and uses – a central question remains: What made their manifestations possible?

The interview and archive data suggest that two influencing factors were outside the control of the advocates; Part 3 begins with a discussion of these. But as neither of these factors is new or surprising, a closer investigation into other influencing factors is also warranted. An analysis of these other factors reveals considerable differences between the three monuments in terms of the relative placement of the three parks, the social positions of the advocates and, especially, the creative and particular agency of these advocates to succeed, against long odds, in creating and installing the monuments. The final chapter postulates a politics of visibility, of relevance to these monuments but also of relevance beyond them, to contemporary debates about public space, social justice and public memory.

Continuousness of the Issue

As the narratives in Part 1 demonstrate, the advocates for each monument were deeply embedded in their communities. The singular climate of the groups' working environment also powerfully influenced the projects. And this climate was contradictory, but proved consequential: there was a tension between how residents of Vancouver and Canada wished to imagine themselves, as strongly opposed to violence against women, and there was the evidence that suggested a continuous and disturbing social tolerance of violence against women. Indeed, just as Chivallon's work in Bristol revealed a wounded memory process regarding slavery, so, too, here in Vancouver, the long shadow of colonialism's violent and exploitative history appears to have "wounded" people's capacity to face what was apparently happening to women on the Downtown Eastside.

The ways colonialism, racism and exploitation have run through the Canadian fabric since European contact are extensively discussed across government, public policy and academic circles. But adding gender to these analyses produces very disturbing results, ones that could, as in Bristol, "wound" the ability to build coherent and robust social memory. For example, in a recent report, Amnesty International noted that Aboriginal women aged 25 to 44 are five times more likely than other Canadian women of the same age to die of violence. Noting that indigenous women are often placed in conditions of "extreme poverty, homelessness and prostitution" (not least as a result of government policies), Amnesty International further observed a profound failing among Canadian officials and police "to protect aboriginal women from violent attacks" and "ignoring the acts when they occur" (Amnesty International Canada 2004). Over 500 Aboriginal women had gone missing or been found murdered in 30 years as of that date, but, hauntingly, Amnesty International remarked that "the people who carry out the violent acts believe that societal indifference to aboriginal women will allow them to escape justice." This seems to have been informed in no small part by what was happening, and not happening, on the streets of the Downtown Eastside.

By the mid-1990s, it was obvious that there was an imbalance between how December 6 functioned at a national and local level to highlight issues of violence against women and the near absence of attention to violence against the women of Vancouver's Downtown Eastside. Many people had noted this discrepancy – including those associated with all three monument projects – and many were aware of each other's work, even if they were not directly collaborating.

Early on, a number of Downtown Eastside agencies criticized the *Marker of Change* project, arguing that donations should support direct services rather than the establishment of a monument. Many others knew about local levels of violence in the Downtown Eastside, at least anecdotally. Articles, poems and short ads appeared regularly in the Carnegie *Newsletter*, highlighting the daily realities of assault and murder, and the toll these took on the friends and families of the victims. Police reports were filed when bodies were found; others were submitted when families insisted that their loved ones were missing. But the fact of the matter is, even at this writing several years on, the extent and particulars of the violence against women of the Downtown Eastside are still unclear, and there seem to be myriad obstacles to ever achieving a complete picture of the truth.

What is clear is that the topic at the centre of the monument projects – violence – has been an ever-constant presence during the time the monuments' proponents have been active. It has never been possible for those opposing the monuments to claim that the issue of violence – especially against women – is not a significant social phenomenon. There have been attempts to argue its prevalence (as in the letters and editorials offered in the summer of 1993), attempts to nullify its gender dimensions and certainly attempts to reduce or dismiss the problem as one of private concern, as is often the case with, for example, domestic disputes. But never has anyone been able to deny that violence, particularly against women, is evident and endemic.

In 1991 (the year the December 6 private member's bill was unanimously adopted by Parliament), the Subcommittee on the Status of Women of the Standing Committee on Health and Welfare published the results of its study on abuse. These results indicated that "a woman is hit by her husband or partner an average of 30 times before she calls the police." In 1994, Statistics Canada corroborated this finding in its one-time-only Violence Against Women Survey, stating that only 14 percent of all violent incidents against Canadian women were reported to police. It went on to state that 51 percent of Canadian women surveyed had experienced at least one incident of physical or sexual violence since the age of 16, and almost 45 percent of Canadian women had experienced violence by men known to them (dates, boyfriends, marital partners, friends, family or neighbours).[1]

1 http://www.statcan.gc.ca, search "Violence Against Women Survey."

Locally, the situation was, if anything, worse than the national research suggested. Using newspaper files and police records, three reporters from the *Vancouver Sun* determined that, as of the end of 2001, "since 1980, there have been at least 65 confirmed homicides of women who either worked in BC's sex trade or were vulnerable to predators because they used drugs, worked as exotic dancers, lived on the streets, or hitched rides. These are cases where bodies were found and there was clear evidence of foul play ... [and] 40 by the *Sun's* tally – remain unsolved" (Culbert, Kines and Bolan 2001).

Between 1989 and 1997 (the years between the conception of *Marker of Change* and the installation of the first two monuments), 17 local women were killed, their killers never caught, and 19 women went missing. Possibly these numbers were known by someone – perhaps by the police or one of the Downtown Eastside agencies. But it is equally possible that they were known by no one at all. It was only when Wayne Leng's personal appeal went up on telephone polls in the area, asking people to contact him at his home phone number, that the tally became formally public for the first time. Still the carnage did not stop – since 1998, several local women have been murdered and at least 19 women have been reported missing.

The point here is that everyday realities ensured that the issue of violence against women was constantly, painfully present. There was a pressing need for these monuments to recognize and commemorate the *sasha* (the living-dead) rather than the *zamani* (the revered, generalized ancestors). A number of social/performative forms were employed to keep this issue in the public eye, the Valentine's Day marches being the most visible and explicit. But these episodic, ritualistic forms are, by their nature, primarily for emphasis; they do not function as continuous witnesses encouraging counter-hegemonic understandings.

These realities were confronted by the "impression management" tactics that larger social groupings sought to employ so that Canada, Vancouver and the Downtown Eastside itself would be portrayed in a better light. By all accounts, the Montréal Massacre stunned the nation. Over and over people said, "But that [kind of violence] doesn't happen

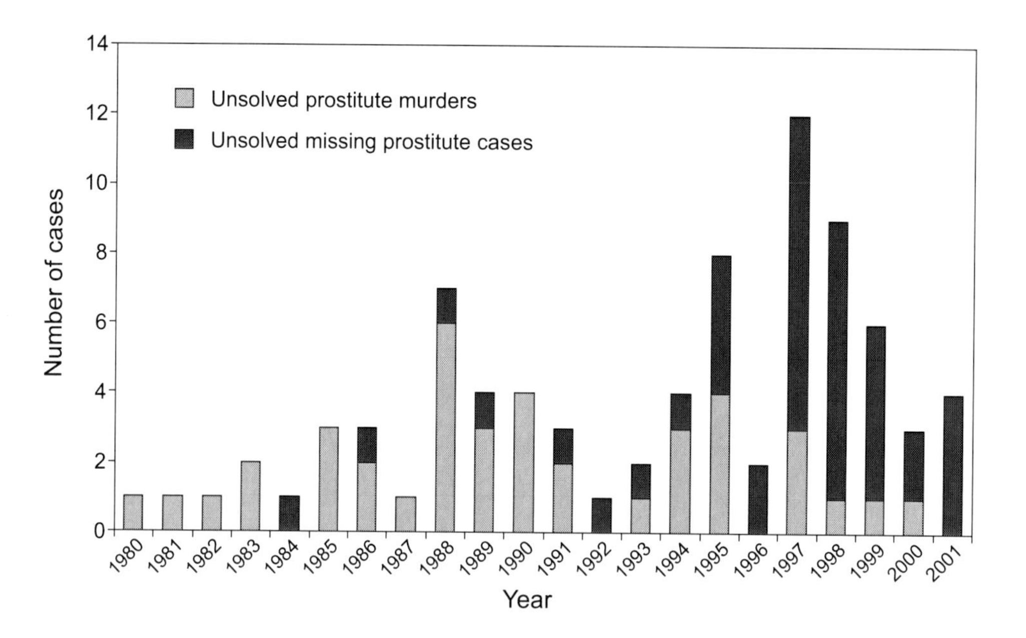

The number of women who have gone missing or been found murdered in this small area caused one local policeman and a local criminologist to conclude that at least one serial killer was likely at work. | *Source: Tara Adair, based on* Vancouver Sun *archived article series (2001)*

in Canada." In fact, it hadn't happened before to a group of white women, in a university. The need to face such violence as a reality of Canadian experience, however, definitely produced splits. In an extended interview following the Massacre, the CBC's most senior national correspondent at the time, Barbara Frum, repeatedly altered the words of those she was interviewing so as to describe the attack as the work of a lunatic gunman and to "erase" references to the gunman's particular desire to kill women. This, despite what was written in the gunman's suicide note, and despite the fact that witnesses report that, prior to shooting, the gunman stated his explicit desire was to kill women because they were women (Bradley 1995).

As evidenced by the *Marker of Change* inscription controversies of 1993 and 1994, facing the "male" aspects of violence against women was an enormously difficult step to take. Yet the national context for this conversation was set, laid out by the actions, legislation, foundations and surveys detailed above, as well as by numerous small memorials referencing the Massacre – in Ontario, Manitoba, Alberta and elsewhere in BC. Key in this regard was the move on the part of many engineering schools to focus on hiring and mentoring women into engineering. Once the decision had been made to locate the national monument project in Vancouver, of course, the city itself became involved in how the role of men in violence against women would be portrayed.

When it was first approached by monument advocates, the Vancouver Park Board did have in place a policy on park monuments, and indeed, there were already nearly 100 memorials in public city parks. But the board's ambivalence about *Marker of Change* can be read in its requirement that the Monument Project Committee prove it represented the wishes of a constituency and that it advocated for an underused space versus a more salubrious park, and in the "out" provided by a board spokesperson in the form of the need for the board's final approval. As the story unfolded, of course, the Park Board did the courageous thing, finally approving the monument, but it also withheld support when it could have given it.

For example, in 1996, advocates for both *Marker of Change* and a proposed Vancouver AIDS monument (whose advocates began organizing in 1994) approached the City for public art funding. Both were turned down. Bryan Newson, then the City's public art programme manager, argued that "the public art program was never designed to deal with projects from special-interest groups." (Note that this appears to imply that people who oppose male violence against women, or who see themselves as affected by AIDS, constitute a "special-interest group." Needless to say, considering the demographic distribution of both of these phenomena, this is a remarkable conclusion). Newson went on to say that "[*Marker of Change*] was accepted on the understanding that no civic funds are required for the art's fabrication, installation, or maintenance." In that year, Newson claimed that the upcoming review of public art policy might "broaden its definition of community to include communities of interest in addition to neighbourhood communities," but when the new policy was revealed, it underplayed this concern in favour of other, more limiting policies:

> The Vancouver Park Board wants to put the emphasis on beauty, not controversy, when it comes to approving new monuments ... The board has quietly brought in a more rigorous policy that could be used to oppose future monuments that might antagonize the public the way the Women's Monument [*Marker of Change*] and the AIDS Memorial have.
>
> ... The board's stricter guidelines ... say memorials should honour their subjects "and will not purposely create antagonism or cause distress" ... The policy also says new monuments must enhance the design of a park. As well, the park commissioners have made it clear they favour "living memorials," such as trees, over monuments made of concrete or steel. (Todd 1998)

For their part, the *Marker of Change* organizers realized how their monument fit into this process. Mused one organizer:

> Because now they have a memorials process, you know. Where they have to go through a whole community process. We were a guinea pig for them. This project raised issues that they hadn't quite had to deal with. They hadn't had a project of this level of controversy. They decided to restrict things further, and put into place this whole other bureaucratic process that actually makes it more difficult for people. And I understand the rationale for putting those in place, because they may get so many demands of that use of public space.
>
> ... I think our memorial, though ... I'm sure it must have taught them to have more trust, 'cause it's turned out incredibly well. There was tremendous fear of what kind of art form would be chosen ... and that's extremely unusual, to say, "Yes, you can have public space" without knowing what the art form will look like. And we [said] "We can't figure out the art form. We can't have a jurying process until they [the artists] know what land they're putting it into" ... *the space was as important as the art work.* So, [it was] impossible for them, but – somehow they did it. (my emphasis)

In the end, many civic agencies and civic employees were among the thousands of donors, so personal levels of support were very different from policy levels of support in regard to the building of the monument. In other areas, the city made its priorities more evident. *Marker of Change* was dedicated in December 1997. One month later, an advocate for sex-trade workers, Jamie Lee Hamilton, dropped 67 pairs of stiletto heels on the steps of City Hall – one pair, she claimed, for each Downtown Eastside sex trade worker missing or murdered in an unsolved case. At the time, Vancouver was undergoing a rash of so-called home invasions that were predominantly crimes against property in the city's more affluent west side. For those cases, the City had quickly offered a $100,000 cash reward for information; no such award was forthcoming for the cases noted by Hamilton. Indeed, it took nearly a year and a half for the city to offer such a reward in relation to the missing women of the Downtown Eastside.

In that year and a half, several things changed. The efforts to trace Sarah de Vries resulted in the collation of details of 31 missing women. Wayne Leng (and others) used this information to confront the police and go to the newspapers. Leng also established a website (www.missingpeople.net), with links to similar sites. Something here drew the attention of "big" media – including the screenwriters for *Da Vinci's Inquest* (a nationally popular TV crime series based on an actual Vancouver coroner), a prominent regional features writer (Daniel Wood) and the crew of the television crime show *America's Most Wanted*. A feature documentary based on the making of *Marker of Change* (called *Marker of Change: The Story of the Women's Monument*) was completed; in it, several powerful sequences highlight the issue of violence against women in the Downtown Eastside. When the documentary was shown nationally, in areas that had supported *Marker of Change*, national interest in the neighbourhood became markedly more informed. Other pieces also increased popular understanding of Vancouver's "missing women," if often in a sensationalized manner; for example, *America's Most Wanted* broadcast its segment to 14 million viewers worldwide. In 1997 alone, an additional nine women went missing: at this point, even the local press began to take notice. All of these media lenses concentrated a great deal of pressure on the city of Vancouver to look like it was "doing something."

For the first time, the police stopped treating the women's disappearances as isolated events, and began to explore the possibility of patterns. The lone police officer working on the cases was joined by two homicide detectives over the next year, as well as an expert in geographic and psychological profiles, a veteran beat officer, a lay person, a private detective firm, and two officers from the missing persons division. The City mounted another $100,000 reward, and launched a website and widely distributed a poster that grouped 31 of the missing women together. The police began to regularly review the "Bad Date Book" put out by the sex trade workers every week to warn women about known bad tricks. In addition, two constables developed a program that notifies "johns" (customers of prostitutes) that they have been seen soliciting and can no longer solicit anonymously. Since these changes were implemented, the disappearance rate has dropped, but still nearly two dozen women in the Vancouver area alone were reported missing in the two years after Hamilton dropped those stilettos on the doorsteps of City Hall.

Over time, however, the level of police response once again diminished. The numbers of specialists and the pool of police expertise dwindled as only four women were found – two alive and two dead – even though more women were reported missing in the same time period, and more were killed. Vancouver police continue to have a "perception" problem in this area in that, arguably, the world's expert on geographic profiling of criminals, Kim Rossmo (who developed his work in Vancouver), has issues serious enough with the department to have resulted in litigation and his relocation to a police department in the United States. Given the legal ramifications, Rossmo is extremely circumspect about what he says regarding the police investigations and *lack* of investigation that has characterized most of this period. But even his careful pronouncements create an impression that Vancouver's police are culpable and possibly even criminally negligent regarding the ways in which Vancouver's Downtown Eastside women suffer violence.[2] Calls have been recently made for a Royal Commission of Inquiry into the long delay in police action on these files. And indeed, accusations of murder against

2 http://www.missingpeople.net, search "Kim Rossmo."

particular (and as yet unarrested) people were levelled during the course of this research by various community residents.

In 2001, things changed. Local police were joined by the RCMP; for a time, their combined staff totalled over 100 experts and officers. In February 2002, the first crime scene (the Pickton farm), the first arrest and the first charges were made public in some 18 years of unsolved missing-persons cases. Police remained at the crime scene and a related property for over a year, and found remains of 16 of the missing women, with some expectations that more remains would be found. The details of this investigation are staggering: 600,000 exhibits were seized by police, and 200,000 DNA samples were processed. Indeed, the Pickton case became the largest crime investigation of its kind in Canadian history, and the largest serial killing investigation ever in this country, costing over a seven-year period nearly $200 million – and it is still not over (Cameron 2007). Although some elements are still under a publication ban pending appeals, we do know that of the 65 women officially identified as missing, Pickton has been charged with killing 26 of them; so far there has been sufficiently robust DNA evidence to proceed with a trial in only six of these cases. Meanwhile, male violence against women continues to claim women's lives.

These developments underscore that the issue of violence against women, especially violence done by men, is part of the fabric of daily life. If a country, city or neighbourhood wishes to portray itself as outraged by this, or as committed to doing something about violence, it cannot afford to pretend it is a private concern best handled between individuals. The three monuments considered here are involved in the production of public responses, and serve as a kind of indictment of the inadequacy of previous approaches to violence. Once publicness (or "collective responsibility") has been invoked – whether imaginatively, discursively or physically – geographic scales of analysis can come into play. Vancouver's responses are watched elsewhere: the reasons for Rossmo's demotion in Vancouver and subsequent relocation becomes news in Washington, DC, and now Texas; the developments in the case of the missing women show up on New Zealand television, on the Canadian national news, in murderers' living rooms and, bizarrely, as part of a (very short-lived) ad campaign for Canada's national newspaper the *Globe and Mail.*

Acknowledging the Unseen

In the accounts of the advocates, in addition to the recognition of ongoing violence, there was also open acknowledgement of another continuity, though one that cannot be easily measured or analyzed. In a review of extensive field notes and interview data, it is striking how often something unseen or immeasurable was credited as having a powerful influence on material reality. While this unseen element was never discussed using the language of religion, it was described variously using words of spiritual significance: "presence," "visitation," "protection."

For *Marker of Change*, both the artist Beth Alber and Councillor Chiavario felt there was something about this project that allowed it to overcome obstacles as diverse as media storms and city strikes ("this monument wants to be in the world"). In addition, *Marker* advocates admitted to being so impressed by the power of First Nations rituals when they were used in meetings in the Downtown Eastside or on the Valentine's Day marches that they learned to embrace ritual themselves. During preparations for the monument, besides a self-reported sense of feeling awkward, the committee asked for help to create rituals both to manifest the monument and to assist in fundraising. In the uses of the monument now, certain rituals are evident, including particular paths of approach, the kinds of gifts offered, patterns of washing and tending the site, and times of visitation. At the groundbreaking, one of the elders appealed to the "spirit of Mother Earth." Additionally, the advocates often commented that the monument itself somehow softened "hard" men, and invited a different kind of interaction. But most of the commentary centred on the day of the unveiling:

> How many December 6th's have we stood out in the pouring rain? ... and the weather was absolutely crappy up until the [moment] ... That one day, it was so crisp and gorgeous ... That happened a lot, though. We would go there, and the clouds would part, you know? I'm serious. They really do ... Yes, well especially when Suzanne came!
>
> Oh, I mean, I'm not religious, but that was almost a religious experience ... There was something else there. There was another whole realm of experience and emotion. A level of emotion and

> connection to why the monument was there, that I hadn't yet let myself understand ... And in particular, Monsieur X. Because his wife couldn't come. She just hadn't, you know, gone through the grieving process in the same way he had. His son couldn't even say his sister's name. And it just opened my eyes to that.

Others continued: "[There was] ... a strong power there ... it was very, very obvious that the spirits of the 14 women were there. It was almost, for me, palpable. And then, especially crossing [the park towards the monument]. It was like the force or energy got stronger and stronger as we got closer." "I've had feedback from people who came to the unveiling saying 'That was the most powerful thing I've ever been to.'"

This energy might have been due in part to the excitement of the unveiling, but an art worker in the area claimed there is anyway "a kind of protection" over the Downtown Eastside. She characterizes it as mostly a spiritual protection, but says it takes many forms. "There is a kind of super-immunity down here," she said. Her words resonate with me: in my meetings with the community residents' group, in an area of such poverty and ill health, I was stunned how many hold cancer or chronic illnesses at bay, for years. When asked why she personally was here, working on the street with art, she explained, "It was literally a call. I went to that corner and I just felt called. I have no idea why."

My most sustained periods of attendance at the boulder have all been as a witness to rituals, in which the presence of the "spirit world" is openly and easily both acknowledged and assumed by participants. Speaking begins and ends with references to ancestors and pre- (or anti-) modernist identities ("I come from the people of the rivers," "brother and sister creatures" and "all my relations"). The natural world is welcomed and thanked – the wind, the sun, the eagles. And of course, this monument is about grief – a state in which one is particularly open to experiencing a spiritual sensibility.

But while the advocates and supporters of the boulder may express this openness to a spirit world, obviously others, such as the man who power walked his way through the ceremony there, find this usage of the space troublesome (it's a pathway; *therefore* it cannot be a site for a ceremony). What must be remembered about the boulder, unlike the other monuments, is that it functions as the only gravesite for many missing women. When added to the fact that the site is in a gloriously landscaped waterfront park alongside the one pedestrian beach access point for several miles in either direction, conflict is almost certain. The placement of the boulder both highlights and exacerbates the porousness of the boundary between the secular and the unseen, or "spirit world." The open vistas, the presence of animals, the grass, marsh, beach and wetland environments, the quiet (relative to the streets of the Downtown Eastside) – all contribute to a sense of the spiritual. But here sacredness must be enacted next to the mundane and the disrespectful; a sense of spiritual peace can become a matter of (extremely) temporary grace.

There is another complexity here, too. The Downtown Eastside is a profoundly multicultural neighbourhood. Among the most highly nuanced cultural performances are those around expressions of grief and anger. For any monument to function successfully within the diversity of this setting, therefore, its very form would have to accommodate multiple approaches. How one shows respect in this environment depends on the context. Determining the order of invitation among a series of elders, when to use names and when not to use them, how to move and when to speak – all tempered with the desire to acknowledge the unseen – is an extraordinarily subtle endeavour. As one elder commented, "We're all learning here. Many of us are learning our traditions for the first time."

These sorts of attempts to honour relationships over cultural forms (and the challenge of figuring out how to do that) also play a role in how the presence of the unseen is acknowledged at *Standing with Courage, Strength and Pride*. But here, the size of Oppenheimer Park (and the placement of the pole deep within it) and the park's continuous use by neighbourhood inhabitants seem to make multiple readings of the space more natural. The presence of eagles is noted often; eagles offered great comfort to people when they came and stood for hours in the trees during the months of carving, and continue to offer solace when they soar overhead during the Valentine's Day marches. One year, we counted nine eagles in the sky above the marchers. A woman next to me said, "Things will change this year." As it happens, that Valentine's Day march was the march before the Pickton arrest.

As the aforementioned interview with Julia indicates, every action in a material world can be understood as related to an unseen dimension. As she understands it, the relative who danced with a mask while he was addicted did harm – made her ill, caused a death, troubled a fetus. And the mother of one woman implored us to allow her murdered daughter, and the others, to be able to touch the living that February 14, which could be done, she felt, if we could walk without anger. The site of the pole then, as the site most *inhabited* in comparison to the others, appears to be the place where the presence of the unseen is the most presumed, and where the unseen is most credited with agency in the affairs of the material world.

The realities of violence and the support of the unseen cannot on their own account for the highly unusual establishment of these three monuments, so closely spaced in time and location. This is especially true when we consider the fact that the advocates worked only minimally together (Burk 2005). To accomplish a feat like this, something usually only the socially powerful can do, the advocates had to work with remarkable effectiveness and credibility in the imaginative, discursive and physical domains. A second map of the area (see p. 91) gives some insight into what might underlie the advocates' efficacy.

The second map reinforces that the differences between how a resident inhabits and how an outsider frames the Downtown Eastside offer a glimpse into a kind of re-centring of the neighbourhood – one that is removed from the perspective of the visitor and more closely reflects the tactile engagement of the resident (Sasaki 2000). These differences may be subtle: during my research, for example, I noted just a slight change in the personal distances residents assumed in the different venues, activities were more focused or more diffuse and the degree of interaction varied, as in the exchange of names versus simple waves or nods of acknowledgement. The back alleyways of the Downtown Eastside also function as a kind of liminal zone, episodically both private and public (depending on what people are doing), even though activities were always in plain view.

There is a prolific artistic energy in the Downtown Eastside. Indeed, the Downtown Eastside is a vital and even (until very recently) quite stable neighbourhood relative to other neighbourhoods in Vancouver. It boasts a string of victories that would delight most communities as evidence of local initiative, and there is a vibrancy to public life unparalleled in other parts of the city. Entertainments in other parts of the city are frequently purchased (in the form of theatre, movies and concerts, for example) or privately created (like backyard barbecues); in the Downtown Eastside, things work differently. Community markets, the Interurban art project, Carnegie classes, weekly street events and a constant series of public fora and lectures, as well as the established spots for "hanging out," help ensure that entertainment in the Downtown Eastside retains to a large extent a community focus. Recently, the Footprints Project, an extensive series of street mosaics and accompanying banners designed and executed by community artists and Downtown Eastside residents, has been installed across the sidewalks and streets of the neighbourhood. The artwork depicts the area's vibrant history, and is described in a detailed and modest guidebook.[3] Much of life is lived very publicly in the Downtown Eastside (not least because the single-room-occupancy hotels that serve as places to sleep are so noxious), which has advantages and disadvantages. On the one hand, it allows residents to interact extensively and frequently with each other, making the Downtown Eastside akin to a village rather than a concentrated urban environment. On the other hand, those neighbourhood encounters, as well as many other exchanges of a more private nature, are virtually continually accessible to the media, or in view of car and cross-town public-transit passengers. Thus, those living in the Downtown Eastside are almost continuously being framed by outsiders.

Indeed, the Downtown Eastside must be far and away the most photographed and videotaped neighbourhood in the city. (Although "beauty shots" of Vancouver's other environments are used in promotional media, as a neighbourhood, the Downtown Eastside is the most frequently profiled in the news and has appeared on three highly popular national television shows.) As the Downtown Eastside is so continuously used as a source of media narratives about Vancouver, it is important to consider what stories these narratives tell, and how, alternatively, the same events and experiences feature in the lived lives of residents.

3 *Old Vancouver*, undated. Copy in possession of the author.

While census figures repeatedly indicate that most of the approximately 16,500 people in the Downtown Eastside are men, most residents are not addicts, not missing, not Aboriginal, not sex trade workers and not criminals. Most are simply poor (67 percent), aging (the area has nearly twice as many seniors and half as many children and youth as Vancouver overall) and alone (more than half of Downtown Eastside residents live by themselves).[4] But in the media narratives, the identity *imposed* on the Downtown Eastside is already set: certain areas and alleyways along four or five particular blocks are used relentlessly to anchor the imagery and media texts generated mostly by outsiders. These framed views (the discursive domain) reaffirm certain preconceived notions for all those who don't live in the area.

When experience of spaces exists only imaginatively and discursively, without the truth-grounding of empirical experience, inappropriate conclusions can become naturalized. Powerful effects can ensue as the *imagined* city becomes the discursive and finally the physical city. It is particularly problematic when these assumptions are visited not only on city spaces but on the bodies within them. Sherene Razack's (2000) work is useful here: her analysis is based on legal interpretations of sentencing for a particularly brutal crime in which the spaces of another Canadian city (including the habitual living environments of the perpetrators and the victim) appeared to feature in the judge's decisions. Her analysis rests on a supposition that cities are imagined, and thus discursively and physically rendered into a continuum of zones stretching from those of rationality/organization to irrationality/disorganization. *Rational zones* are ordered and characterized by disciplining procedures and artifacts; in contrast, *irrational zones* are areas cast as disorganized, racialized, sexualized and imbricated with an implicit acceptance (and prevalence) of violence. In Razack's formation, rational *persons* (subjects based on liberal, Eurocentric ideas of alienated property relations that mandate that subjects be constructed as abstracted from place) enjoy a kind of coherence and immunity from place unless they find themselves in irrational zones; in these environments, they can be tainted and

4 www.city.vancouver.bc.ca/dtes.

"overwhelmed," and thus excused for indulging in irrational behaviour. By contrast, subjects whose habitual living environments are deemed to fall within a city's irrational zones lack the ability to become rational. In her legal judgment, Razack found that an irrational subject *cannot become rational, no matter where s/he is.* Rational subjects, however, are presumed, almost even expected, to behave irrationally in a zone of disorganization. In this way, the imagined subject, reified discursively, also becomes physically aligned with particular characteristics.

Razack's reasoning is uncomfortably relevant in our study of the places and people of the Downtown Eastside. Despite the vibrancy of the neighbourhood, the area identified as the Downtown Eastside would be the primary area designated – by many in contemporary Vancouver, including residents of the Downtown Eastside themselves – as embodying the characteristics most associated with a zone of disorganization. To date, these characteristics or impressions of them have been externally produced (Anderson 1991; Blomley 1997), reified in policy (Sommers 2002), framed rather relentlessly by media (*America's Most Wanted* 1999; *BCTV News* 2006-07; Carson 2002; *CBC* 1998-2003; Gill 2001; Kovanic and McCrea 1999; Kovanic and Johnson 2000), autobiographically recorded in residents' accounts of lived experience (*Carnegie Newsletters,* 1986-present; Grove 2000; Osborn 1989, 1995) or identified in statistical terms for health, housing and other demographic measures.[5] The "hundred block rock," or the central intersection of "Pain and Wastings" (Main and Hastings) are vernacular terms of considerable meaning to local residents. Also, as evident in repeated field notes over several years, neighbourhood legitimacy is drawn in no small part from displaying both bodily and in conversation a personal legacy of addiction or of survival of addiction and/or violence. Nevertheless, the media's preoccupation with labelling the area as "the worst" in terms of HIV/drug addiction (in relation, variously, to North America, Europe, the industrialized world) and its identification in the Canadian public imagination with the violence associated with the missing women have now fully blended the realities of its streets with heavily sexualized and racialized impressions. Violence,

5 www.city.vancouver.bc.ca/dtes.

disease, racial otherness (in the form of female Aboriginal sex trade workers) and dependency are thus overtly displayed in the neighbourhood *but they are also* invented associations.

So the area of the Downtown Eastside *is* cast as a zone of disorganization – both internally and externally – and it *is* saturated with the prevalence of violence. But the microgeographies suggested by the second map, which displays different kinds of publicness and the different kinds of inhabitance of the neighbourhood, show that the Downtown Eastside cannot be viewed as a uniform zone, even within the relatively small area circumscribed by the three monuments. There appears to be some contestation about visibility itself.

This may have something to do with the proximity of the Downtown Eastside to Vancouver's rapidly gentrifying central city. Economic and socially hegemonic forces stand to gain valuable real estate by insisting that the Downtown Eastside be regarded as residual or disposable space (recall the case of Africville, cast as so obviously miserable that obliteration was seen as a form of rescue). This wouldn't be a new tactic: the neighbourhood and its residents have for over a century been identified in the public imagination with the socially abject, thus permitting "tainted individuals and the spaces they inhabit [to be] removed from the field of local moral concern" (Woolford 2001, 29, regarding HIV/AIDS; see also Anderson 1991 regarding Chinese residents; Sommers 2002 regarding failed masculinity). It is in the interests of some, therefore, to continue to cast those in Vancouver's Downtown Eastside as irrational inhabitants of violent, racialized and sexualized spaces in need of "fixing," as per an extensive four-part feature series of Canada's national newspaper (Our Nation's Slum ..., *Globe and Mail* 2009). Given this, the celebration and enrichment of collective memories and linking these to particular places in the Downtown Eastside is profoundly problematic for hegemonic forces. For them, the memories in such areas are better obscured than commemorated. Perhaps, as Sennett noted in the 1960s, such forces would prefer to eliminate what disturbs (the troublesome bodies and memories) and pre-empt *actual* participation in community with instead an *imagined* state of belonging, and a relegating to history of all those who no longer fit the desired type of resident. But such a strategy works only if the images and associations of a zone of disorganization can be made to stick as the only images and associations of an area. While the media treatment has worked to reinforce the framing around the Downtown Eastside, the monument advocates have understood this, and have made use of it, both to invite (and to elude) scrutiny, participation and witnessing; their efforts reveal a profound understanding of these lived microgeographies of the neighbourhood as well as a marked savvy about how to use (and not use) media.

This is clearly demonstrated by the varied publicness of the three monument sites. Originally, all three sites were part of the extensive fishing grounds and wetlands shared by a number of First Nations, their camps open to the great waters of Burrard Inlet. But post-European contact, the sites have been developed quite differently. Of the three, CRAB Park is the newest park. Reclaimed from a scarred landscape littered with the detritus of Vancouver's earliest industries, CRAB was deemed a park space only in the last 30 years, though it is now officially designated green space on the modern urban grid. Thornton was established at the turn of the last century as an Edwardian gateway to the west, introducing Vancouver as the final destination of the Canadian Pacific Railway. The park was a prime example of civic refashioning from its mass importation of soil to the laying out of marked boundaries, paths and plantings. Though it has gone through periods of underuse, Thornton has retained its distinct status as a civic property for nearly a century. In contrast, Oppenheimer was a central gathering place possibly for thousands of years before it became a more formalized village square and, as the Powell Street grounds, a centre for the Japanese community in Vancouver's infancy. Before the creation of CRAB, it would have been one of the public areas closest to the water's edge and thus the world of the sea.

These parks have featured significantly in the Downtown Eastside neighbourhood and, indeed, in the lives of many of those who would feel in some sense represented by the monuments. For example, in the recent past, it was in the battles for the establishment of CRAB Park and the liberation of Oppenheimer park from drug dealing, in addition to the fights for a community centre, increased housing and a community garden, that residents were able to establish this area of the city *as* a neighbourhood and not simply as another skid row.

By intentionally working with such resonances in these sites in the imaginative and discursive realms of public space, the monument advocates each expanded the claims of community at the physical level. For example, the *Marker of Change* promotional literature talks about the placement of the monument in a "heritage site ... among some of the oldest trees of the city." The proponents did not wish to usurp the land and replace it with something new; they understood and appreciated that the park had existed, albeit somewhat neglected, almost free from development for nearly 100 years. Given that the park's designation had been preserved even during the Expo '86 building boom, it seemed likely that *Marker of Change* could very well be in place decades from now. Also, the placement of the monument helped enhance its message: working with the notion that a monument in Thornton would reflect the park's spatial identity as "of the city but belonging to many neighbourhoods," advocates hoped to draw attention to the fact that many spaces were dangerous for women.

Placement matters for the other sites as well. At CRAB Park, the boulder sits very close to the shore, but its textual inscription faces the city alongside the one major asphalt pathway in the park, ensuring its visibility to passersby. The boulder's dedication in July 1997, as part of the 10-year celebration of CRAB Park, suggests a sensitivity to the fact that CRAB is most used physically in the summer, though it is significant imaginatively and discursively in the narrative of the neighbourhood in all seasons. This date also suggests that the particular, awful grief borne by the families of the murdered women did not need to take place in isolation; rather, it was acknowledged amid a number of activities celebrating community, resilience, festivity and connection. Finally, the boulder's CRAB Park site places it squarely within a broader landscape of conversations about property relations in the Downtown Eastside. Thus, though the inspiration for the monument was generated by a very small group, the boulder was deliberately placed and initiated into the neighbourhood through activities and rituals that identified it in public as belonging to the inhabitants of the Downtown Eastside.

The monument in Oppenheimer Park suggests similar circumstances, though here there are much deeper resonances in the site. The 100 years of local baseball games, the multiple annual festivals and the myriad daily activities at the site reveal this as a busy social space. The topography of the city block that is Oppenheimer is quite varied – flat and dusty in the northwest corner of the baseball field but characterized by slight rises and dips on the eastern side. The pole sits in one of the dips on this eastern side, well inside the park and angled, close to a pathway. Most of those who pass through the park walk right beside it. The pole faces west/ northwest, towards the Carnegie outreach buildings, where there is a community kitchen, an office and an activities room; the Raven and the other figures look out over the park.

The use of Oppenheimer as not only the imagined and negotiated site for the pole but also the physical site of its construction indicates a kind of brilliant insight into the non-linear sensibility towards time that characterizes exchanges in the Downtown Eastside. Daily activity in this place for many months allowed those across a wide social spectrum to take part in the monument's creation at all kinds of levels – in imagined, discursive and physical ways. Long-term residents, those just passing through, people's relatives, those in and out of illness or addictive cycles or prison – all were able to participate in ways not possible with the other two monuments. And the public events of the raising, feasting and plaque mounting also extended the circle of those who could be thought of as part of the constituency of support for this monument.

Of the three sites, Oppenheimer is the most socially complex, not only in contemporary use but also in imagined and experienced history of both First Nations and other residents. The cumulative displacements and evictions are likely sharper for Oppenheimer than for the other two locations; there is more recent memory and more human experience of the place to appeal to and/or dislodge. It is interesting then to juxtapose the necessity for the pole *after* the other monuments were installed with the realities of Oppenheimer as a location most vibrantly public in terms of both its continuous use as a space and its centrality to social life in the Downtown Eastside today. Arguably, the three monuments represent different frontiers of claim – against acts of denial and erasure or as a burial of the misconception that the Downtown Eastside is a residual (and therefore disposable) city zone rather than a neighbourhood and a home. Indeed, the monuments blur even the boundaries of the Downtown Eastside, in that they complicate the relationships of place and memory (that is, they draw on elements both "of this place" and beyond).

Arguably, Oppenheimer serves as the central memory site for a sense of community in the Downtown Eastside, since it is a powerful symbolic as well as social site. So it is interesting that this monument references the highest number of dead, not by name, or even by time period, but by clustering together all "our sisters and brothers who have died unnecessarily in the downtown eastside." It thus eloquently includes not only those who have died from poverty, violence, addiction and other contemporary assaults, but also those who have been ravaged by colonialism and the violences of racism and structural exclusions. It is here that the monument is the most personal, created mostly in situ, carved literally by the relations and friends of those who died and by those who were dying themselves. Of the three monuments, this was the one that was installed and feasted by neighbours. It is here that the monument reveals most vividly Evelyn Peters' (2002) work on the multiple spatial scales of identity that urban First Nations people claim are necessary to negotiate for good mental health. This requires a recognition of negotiating one's place *simultaneously* at several different levels of spatial identity: not only with the immediate physical (urban) location, but also in relation to "Mother Earth" and "the community of First Nations" – both those related by blood and by imagination. It would seem there is some evidence of these efforts in the tobacco, sweetgrass, notes and rituals that appear frequently at this site.

CRAB Park's boulder also seems to stand on another symbolic ground that resists efforts of denial, erasure and burial of local memories. Though CRAB is the newest park, its evolution has coincided almost precisely with the incidence of women disappearing. Just as the park itself was won through a spirited campaign that insisted on the reality of a neighbourhood that many dismissed as too transient to count, the boulder itself stands in plain sight to mark an absence that many found too unbelievable to countenance. Both the land and the boulder are thus implacable signals of a resistant presence – even when that is the presence of absence.

Finally, Thornton's recent (pre-monument) history as a kind of discarded city space indicates that it occupied what was essentially a location of denial or suppression. Once it was assigned a central civic importance, after its mid-century use by industry and then by labour and peace activists, it was left to drift into the periphery of civic consciousness. But margins can be powerful places from which to work. The monument and the overt education and fundraising processes its advocates undertook have turned this "park-on-the-periphery" into fertile ground. Moreover, if it was hoped by some that the physical marginality of the park would lessen the visibility of the monument, these hopes have been dashed. Now the monument is one of the first things train and bus travellers encounter on their arrival to Vancouver. It is used for discussion in language classes and with new immigrants to Canada. It is visible daily to thousands of SkyTrain passengers. And finally, even though it is peripheral to the Downtown Eastside, its relative nearness forces intense scrutiny of local violence. Arguably, the discussions and antagonisms spawned by its placement are powerfully linked (in a progressive way) with the unravelling of the conspiracy of silence regarding the so-called missing women. The two-pronged attack on *Marker of Change* – both not to say *that* ("women, murdered by men") and not to say that *here* (in this neighbourhood) was met by the monument committee members as an opportunity to learn and educate others. The commitment to raise money from supporters meant the committee members had to attend to and answer criticism thoughtfully. They had to conduct these conversations through the media and in open meetings. As one advocate noted, the reactions to the documentary film about *Marker of Change* proved the efficacy of the monument's spatial proximity to the streets from which women were going missing: "You know, [e]very time we show the film, the scene everyone remembers first is [the one] talking about the women of the Downtown Eastside."

The sites themselves, then, provided a kind of framing for the activities and intentions of the various groups of advocates. But beyond the resonances in the three sites, there remained the hard work of consolidating community, designing site-appropriate work, and succeeding in actually piercing the screen memories obscuring the realities of violence. So, how did they do it? How did the monument advocates transcend their usual social positions to win access to public space, and to design, install and mark the landscape with permanent expressions of social memory? There are a number of identifiable factors.

Consolidating Claims of Community

Quite reasonably, when something as massive as a monument is being planned for a public city park, there is a process, and the first requirement of this process is to demonstrate a legitimate claim as a community or community of interest. In the case of *Marker of Change*, the Monument Project Committee was asked by the Vancouver Park Board to prove that there was a constituency of people who felt a monument memorializing the Montréal victims was important. Given the national outcry, the December 6 Act of Parliament, the gun legislation movement and hundreds of pieces in film, video, newsprint and broadcast media (see Cultural Memory Group 2006), not to mention the Massacre itself, this seemed to committee members to be an odd request. Still, the Park Board was on new ground and it knew it. It did not want to cede parkland for a non-local monument without a demonstration of a constituency. The Monument Project Committee drew up a list of the kinds of spokespeople they felt would represent, in Bryan Newson's words, both a broad and a local community. The list eventually included statements from all three levels of government; from every female leader of all parties in both houses of Parliament; from local and provincial organizations dealing with battered women, women's counselling, women's shelters, disabilities and women's rights; from a Vancouver-based male counsellor dealing with men's violence; from the College Institute Educators Association; and from the department of Women's Studies at Simon Fraser University. Early on, as well, the committee received donations from numerous organizations and corporations, the immediate assistance of a number of people at Capilano College and letters of support from numerous local community centres and Downtown Eastside organizations. And then there were, over the years, the literally thousands of letters and contacts that came in from all parts of Canada and beyond every time the monument was discussed.

Initially, such a collection of support worked to indicate that a "community of interest" existed. This was enough for the Park Board to begin the process of looking at sites and helping the committee think through a public art process. When the first controversy emerged over the wisdom of using money for a monument rather than direct services, there was considerable conflict between community understood in a proximate, local sense and the more imagined (and geographically not-merely-local) community of monument supporters. Members of the Monument Project Committee were accused of not being feminists, at least not thoughtful/committed/good ones. By the time the controversy over the inscription ignited (in July 1993), the Park Board was already nervous about being linked to such volatile public opinion. At this point, the claim of community had to be reconstituted. It was curious in terms of the inscription controversy how often foes of the monument began to speak of the public as needing education and protection. Was this due to an assumption that passion is best taken out of public life as a means to achieve a sort of even public conversation? For its part, the committee's letters and editorials from that point on used (intentionally?) more personal language: "us," "we," "women and men alike" and, in the inscription itself, "we, their sisters and brothers."

Thus, from mid-1993 through the end of 1994 (after the winning design and site had been made final and public), the *Marker of Change* advocates had to consciously recreate their claim of community – in all three domains of public space. To do so, several things occurred: on a national stage, December 6 and the families associated with the Montréal Massacre had achieved consistent and considerable recognition; through the artistic competition and resulting publicity, the monument had become nationally known and associated with Vancouver; voices of men were raised, both locally and nationally, in support of "collective responsibility" rather than denial on the issue of violence against women; and locally, serious and thoughtful attempts had been made to acknowledge the violence of the neighbourhood within the design of the monument. Taken together, these developments allowed for multiple claims of community to co-exist. Still, however, unbeknownst to the proponents, the funding to complete the project was years away. The funding projects, the in-kind donations (of materials such as the granite and tiles, as well as the services), and the numerous speaking engagements, letters, and conversations from 1994 until the unveiling in 1997 continued to represent opportunities for exponentially enlarging the "community of interest."

The claim of community for the CRAB Park boulder appears to have preceded the monument, in that the boulder's commission and installation roused virtually no opposition or particular ceremony. Instead, it appears that the claim of community was tacitly assumed as a direct result of the embodied, physical occupation of the CRAB Park land. As noted above, the demonstration of community within the Downtown Eastside is palpable, but it is also characterized by diffuse, episodic and charismatic politics. Harsh poverty inflicts havoc on lives and tremendously alters peoples' ability to plan, sustain campaigns and even carry forward visions that remain stable. When the distortions of episodic employment, poverty, addiction, illness and premature death are considered, and the social chaos of constantly shifting funding policy are also factored in, the idea of organizing community in other ways rings false. Perhaps this is why those who sought to install the boulder did not claim community in anything other than a sense of placement. They deliberately chose an art piece that was recognizable as a monumental form and did not require a juried process. The boulder names two imagined communities – "Native Aboriginal women" and "the Downtown Eastside" – but does not qualify them further. Yet the boulder's very obliqueness suggests almost an intimate message, like a note left under the door. It seems to imply an understanding that both those affected by Downtown Eastside violence (friends and families) and those in a position to do something about it (the media and the police) would read this monument correctly. In a sense, it offers a very personal form of communication, one that just happens to be in a public space.

The third monument, however, forges its claims to community in numerous imagined, discursive and physical ways. The very idea of a pole, for example, invokes the many traditions of those peoples who carve and raise poles as markers and as places for the spirits to visit the living. All kinds of First Nations poles are recognizable and even celebrated in

Vancouver's performances of "self" (witness BC tourist brochures, maps, and the commissioning of poles for a number of civic occasions). Further, a pole is in keeping with a keen cultural awareness of the high visibility of First Nations people in the Downtown Eastside; in virtually every meeting or gathering I attended in the neighbourhood (and in any government meetings), there have been verbal, behavioural or cultural references made to First Nations' traditions.

Thus the characterization of Oppenheimer Park by terms generally used for a dwelling ("*our* back yard, *our* living room") suggests a large imagined community that is here but is also vibrantly carried internally and imaginatively. As an art therapist who worked in the neighbourhood for years commented, most times when she works with residents of the Downtown Eastside, they talk of the past, of an earlier time when the world had an order it does not have in the present, a time when there was still a sense of a future. These story-worlds are always populated, and people often say that these long-ago characters act as guides in the present. It is in this sense that Oppenheimer seems full of other spaces, other rooms and other people, right alongside its present-day denizens. Also relevant here is the comment that secured the services of the master carver: "In our tradition, if a Native asks you, you cannot refuse"; thus, it is not only the local (proximate) community at issue here, it is also the imagined community of various nations of indigenous peoples and the respect due to persuasive strangers.

The community of the pole shares discursive aspects of a claim of community in a way similar to the constituencies of the boulder. Along with CRAB Park, Oppenheimer has featured large in the counter narratives of community, stories Downtown Eastside residents tell themselves and others. As with CRAB Park, the physical amenities of Oppenheimer speak to repeated battles for the City's recognition, and repeated claims of property rights.

Physically, of course, the pole was carved in the park, by people familiar with the park and its residents. In this sense, it lays the most grounded *physical* claim of community – truly "done *by* us, not *for* us." The pole raising and feasting were wholly local enterprises – from the cooking of the food to the lifting of the pole. Though all of the monuments were designed as highly accessible and in that sense are all interactive, the pole most fully realizes this aim, in that such poles traditionally are seen as alive, as members of the community that feasts them.

Design Features

The three monuments presented enormous design challenges for artists, who had to take terribly upsetting memories and events and make a place for them, a place that could contain grief and despair but still suggest something beyond them. The best monuments do that – they lead us into the abyss but they do not abandon us there. Interestingly, each of the three monuments used a different approach.

The design of *Marker of Change* involves an extremely simple form – a low-lying circle of 14 benches, encircled in turn by a ring of tiles. The material forms are rooted, and carved with coffin-like detailing from pink tombstone granite. Within the landscaping of the park, the benches are neither visually nor materially overwhelming in the space. There is no single point of domination – either visual or physical – and approach is possible from virtually every angle, by virtually every kind of transportation and every kind of body. Several of the proponents of this monument have said that the design itself somehow seems to possess an agency or ability to alter social relations. Though many worried the monument might actually attract violence, another kind of experience seems to prevail. Several women have noted that before the monument was installed and during its installation, they experienced or knew of unpleasant and hostile interactions with men in the park. But noting the experience of meeting men at the monument, three contributed:

> I think that the monument, talking about murder and women, touches the men that used to be threatening and brings up some other thing in them ... Why they're down at Thornton Park, and living in that area, and have such struggle in their lives, is validated ... And that they've lived a lot of violence. They've witnessed a lot of violence ... And they're softer ... [Once] a young man started talking to me ... he was very quickly into telling me about the abuse he was experiencing in a gay relationship, and all this stuff ... and it was like, wow. Did we ever think, five years ago, that this ... is going to assist some young gay guy in talking to some stranger

"One of the major concerns of the women's monument project was that somehow women would 'take revenge.' Given the opportunity, we'd have, yes, on rearing horses women holding up men's heads or something. Right? There wasn't one submission that aimed in that direction. Not one piece insinuated hatred of men. But what there was, was an incredible expression of grief ... and the sadness that women felt ... And yes, anger, but not anger in that kind of road rage mentality. Anger ... the kind of determination that this must stop."

— Haruko Okano, member of the Art Jury (excerpted from on-camera interview in *Marker of Change* film)

about what he was experiencing? The oppression, and fear, and all that?

The physical design seems at the most fundamental level to be one of inclusion. This has the effect of reinforcing an expansive, rather than a singular, sense of community. And clearly, its ceremonial uses – to commemorate Dr. Bernard Slepian's murder, to acknowledge the events of 9/11 and December 6 and to celebrate International Women's Day – are open appeals across difference. But the design can also be read on a finer scale, and thus work to invoke other kinds of communities as well as considerable intimacy. The very forms, suggestive of fallen bodies, are immediately sobering, and the scars (which suggest both gashes and vulvar shapes), are disquietingly personal. In text, both silence and voice are expressed here – although each bench bears the name of one of the Montréal victims, seven of the benches are uninscribed, to represent the unspeakable, and to represent those who cannot speak. The inscription, which attempts to include all women murdered by men, and all "sisters and brothers" who work against this, appears in seven languages and Braille – another deliberate gesture of inclusion. There are over 6,000 donors listed by name, but there are also those who requested anonymity, for a variety of reasons. But as one organizer noted, "the donor tiles are about individuals putting their names behind something. The courage to do that. When people donate you mustn't ever think it's just money. They *don't do* it unless they care." This point is underscored by the family tiles

Annie St-Arneault

Bien-aimée Annie, Tu as été un cadeau du ciel. De là-haut, continue de nous guider, de nous garder dans l'amour et la paix, dans une société de non-violence pour tous les peuples. Avec amour, nous nous souvenons. Ta famille: Laurette, Bastien, Serge, Sylvain, Lucie St-Arneault.

Beloved Annie, you were a gift from heaven. From up there, continue to guide us, to keep us in love and peace, in a society of non-violence for all peoples. With love, we remember. Your family: Laurette, Bastien, Serge, Sylvain, Lucie St-Arneault.

and the tile for the Downtown Eastside. These are the only tiles that present whole thoughts, and some of them read almost as personally as letters – extraordinary expressions to set with permanence in a public space.

When asked about the design form of the boulder, one of the key organizers said that the good thing about boulders is "that they are from several traditions," identifying several cultural groups that in fact had come as original peoples or in immigration waves into the area. The idea of this material and this form was that it was natural and permanent: the key was "not to change it more than you have to." There are differing accounts of the text's origins, but the effort to be specific about the cause ("in the spirit of the people murdered in the Downtown Eastside") without using personal names reveals an intention to include all the victims of this area, across time and circumstance (for example, victims of alcohol poisoning as well as violence). If you and your family or friends feel included, then you are. Still, as with *Marker of Change,* there is a direct invocation of certain communities within this larger inclusion, specifically "Native Aboriginal women, murdered in the Downtown Eastside." It is perhaps telling that all the text is in English.

The boulder proponents did not engage in a design process. They imagined the monument, secured Park Board permission, and gave the desired text to a professional tombstone-engraving firm, with a contract to have it delivered in time for the summer 10-year anniversary festival. Yet something about the simplicity of the form and the openness of the

Maryse Laganière

Maryse Laganière
Elle n'a vécu
que ce que vit
une rose ...

Maryse Laganière, she lived only
what a rose lived ...

inscription seems to have suggested inclusiveness nevertheless. Kelly White, an Aboriginal member of CRAB-Water for Life Society, is reported in the Park Board minutes as saying: "[This] memorial is crucial for their community and is part of the healing process. It is also intended to develop a strong partnership economically and socially and to celebrate the different communities living together ... A lot of the families have not been able to return to their homes to pay the natural respect of dying. Even though the initial project was for the Aboriginal women in the Downtown Eastside, other nationalities have approached the group to assist them" (Vancouver Park Board, April 28, 1997).

In contrast, the design elements of the pole are more explicit. As Dick Baker noted, the animals represented on the pole are readable by all sorts of peoples and all sorts of communities. Furthermore, especially within Canada, the form of the pole itself is not unknown or startling. The pole here, as identified by all interviewed, is not a house pole or mortuary pole but, rather, a memorial pole. Thus, though it differs considerably from traditional poles, its purpose is not ambiguous.

The ability of each group of advocates to work so deftly with what in practice turned out to be three quite different constituencies and three quite different sites speaks to how they understood their embedding within and accountability to their communities. This understanding is reflected in the methods used and the approaches taken on a journey that was impossible to predict and towards goals that were unlikely to be met. A closer look at how these groups kept themselves going and included others reveals a fierce tenaciousness, certainly, but also a humility.

Geneviève Bergeron

Geneviève était tout l'amour du monde, et avec cet amour, elle allait chercher le meilleur en chacun d'entre nous. Et elle nous rendait meilleur ... May the souvenir of her death help to build a better world, a world of love. Then, only then, Geneviève and her thirteen companions will not have died in vain. And we will be consoled ...

Geneviève was all the love in the world, and with this love, she brought out the best in each of us. And she made us better ...

Street Smarts

One of the strongest impressions that strikes a researcher going through the archive of *Marker of Change* is the constant presence of carefulness in language, whether in fundraising letters, responses to critics or pieces written for publication. When asked about this, all the members of the committee laughed for a long time: "You can't imagine the hours that went into each letter."

It is this level of attention to process – as shown in the archive files, in the committee minutes quoted above and in the "rounds" that accompanied each meeting – that reveals much about the principal methods and skills of this group. The rounds process involved a personal checking-in by each person at the beginning of *every* meeting. As explained to me, "Rounds [is] the act of listening. Way more than half of getting close is listening – people often don't speak articulately. Listening is an act of being in the now. Not being allowed to comment." Though this was enormously time-consuming, several claimed it was the one thing that compelled them to work on the project through thick and thin, all the way to the monument's realization. "It was just fascinating to see how everyone processed everything, and how the dynamics of the groups, and how this feminist thing sort of functioned." Indeed, almost every memory that came up for discussion in the interviews was stated from a personal perspective, but then accompanied by nods, "mmm's," validations and parallel memories. This gave the overall impression that the experience of making this monument was intensely *shared*.

Barbara Daigneault

Barbara, ma toute belle! Pour nous que t'aimons tant, tu es passée dans nos vies telle une étoile filante: le temps d'embrasser nos coeurs et de réclamer dans ton destin, respect, amour et égalité pour toute et tous. Ta mère Henriette, ton frère Jean-Christophe, les Therrien/Daigneault.

Barbara, my beautiful one! For us that love you so much, you have gone through our lives as a shooting star: the time to light up our hearts and claim, in your destiny, respect, love and equality for all. Your mother Henriette, your brother Jean-Christophe, the Therrien/Daigneault family.

This sharing extended even to unknown others. For example, in the controversy about whether or not to withdraw the "by men" wording, there were two illuminating comments: "Don't underestimate the ingenuity of the people submitting," said one. Another added she had been considering withdrawing support for "by men" but "thinks it's because she was afraid. Wants to be forced to change it." Both suggest a trust in a larger social sensibility.

There was also, upon probing, an interesting stance taken towards conflict. Conflict, loss of tempers and huge disagreements were openly acknowledged in the interviews and archives. In discussing the WAVAW letter, and on my noting a slight gap in the minutes, one of the proponents acknowledged that responses to the letter were discussed, and the minutes were edited "[as] we decided we didn't want people to necessarily know how hit we were. And the childish things we said ... They were right, actually, I think, in the end. It's amazing how narrow a concept is. And you try to be so grandiose and broad. And then people just reveal your weaknesses and narrowness to you." Three more continued, discussing a meeting they had had with an influential Aboriginal organizer in the Downtown Eastside:

LIANNE: We were inviting her to speak. And, in order to get to that point, we needed to have some more conversation about what it

Hélène Colgan

Hélène était notre joie et notre fierté son départ prématuré a laissé une plaie que le temps ne peut guérir Clarence et Liliane parents d'Hélène Colgan.

Hélène was our pride and joy. Her premature departure has left a wound that time cannot heal. Clarence and Liliane, parents of Hélène Colgan.

means, and all that. And I remember ... you were saying to her that this monument was for us, for many of us ... it opened our eyes. It raised our awareness about what was happening in our own city. And that we needed this monument as well, to raise that awareness. And that for us personally that was an important aspect of it.

INTERVIEWER: And how was that received?

LIANNE: Well.

CHRIS: She has a big heart.

KIM: Yeah!

CHRIS: I mean, she slams us better than anybody, I feel, but, but you know, I love it when she does it, because she's right.

KIM: Yeah.

CHRIS: She's not doing it for any kind of mean reason.

LIANNE: Mmm. I know.

CHRIS: She's just stating a right point of view within its parameters. Right? I mean, it's not ...

KIM: Yeah.

CHRIS: ... mean, or mean spirited. She has a big heart, and she could see the purpose of what we'd done and how we were ... and she also could see that, somebody from that neighbourhood needed to organize one for that area. And that it couldn't be "the white girls" come in and do it.

VANESSA: Yeah. Mmm.

Maryse Leclair

Celle que nous avons aimée et que nous avons perdue n'est plus où elle était mais elle est toujours et partout où nous sommes. À bientôt Maryse. Papa Maman Madeleine Geneviève Sophie.

The one that we loved and that we lost is no longer where she was, but she is always and everywhere we are. See you soon Maryse. Papa, Mama, Madeleine, Geneviève, Sophie.

This choosing to be open to difference changed organizing in all kinds of ways for the *Marker of Change* group: "Well it's a whole style of inviting other people to organize with you. We learned that you don't say, 'We can't.' Well, if we're inviting other people [we would] bend over backwards. It's just respectful."

Listening and learning from process to this degree, of course, takes considerable amounts of time. Meetings were held at least once weekly for more than seven years, and many of them produced work that had to be done outside of meeting times (phone calls, drafts of writing, paperwork). This would seem to suggest that members of the Monument Project Committee must have been people of leisure, but that is in fact wrong. Some had unconventional schedules (for example, some were musicians, freelance journalists and students), but all worked and attended to family and partners as well. More than one was open about the exhaustion:

> There were definitely times where I was just, like, "Okay. This has gone a lot longer than I planned." But I looked around, and I [thought] "You know, I can't just leave." And I saw the commitment, and I felt that feeling of being close, and needing that. Needing to follow it through [pause]. I mean so many times, I was just so tired, and I just wanted it be over. It frustrated me so much that it wasn't materializing. But there was a feeling and it totally had to do with the people that were involved and the work. And just

Barbara Klucznik

Jak umre napiszcie
na moim grobie:
"Urodzila sie
I uczyla sie ..."

— Barbara Klucznik

When I die, write this on my grave:
"She was born and she was learning ..."

— Barbara Klucznik

thinking "Who else is going to do this?" You know, like there might be someone, but where are they?

The street smarts of this group in part had to do with how they functioned and how they interacted with the world. But given the size of the group (some 40 collective members over the course of the project), there was also an enormous range of skills available at different points. For example, a nuanced use of language (in both French and English) speaks not only to literacy but also to a careful sense of rhetorical register, and to the functions of different kinds of texts. These skills were not equally shared, but the collective process ensured that nothing was issued in the name of the project without it being discussed.

This skill in language was accompanied by a fluency in life experience over a huge range of social areas. Many of the proponents came out of abused backgrounds themselves and/or worked in direct services around these issues. They came from across Canada and drew on both urban and rural life experiences, and had experienced a variety of waged work; some were in their twenties, others decades older. They had health issues, they had deep relationships and they had clearly experienced more than a little death. Few claimed an intimate knowledge of art, though each had a sense that art was something that had the power to move and change people. They knew a variety of ways to work with others, to sit with personal confusion and to organize themselves and others. This

Annie Turcotte

Annie, ton souvenir nous rappellera toujours que la seule chose qui importe est l'émergence de l'amour et l'amitié dans nos relations humaines, afin que l'on puisse espérer vivre dans un monde meilleur. Carmen, René, Donald et Christian.

Annie, your smile will always remind us that the only important thing is the presence of love and friendship in our relationships, so that we may hope and live in a better world. Carmen, René, Donald and Christian.

combination of social experience and social position was profoundly important in bringing the monument to realization.

From the moment the monument was conceived as a national project, one involving a national submission and juried competition, it was clear that this was going to be a massive educational as well as fundraising undertaking. This meant that virtually everything the committee members had to draw on – from work environments, to friends, to travel home to family – became possible sources of support for the project. Workplaces provided venues for events, equipment and locations for conversations and presentations. Friends were asked to take things into their own social worlds and spread the word. They were asked to phone the Park Board, write letters, donate and talk others into supporting the project. Trips to visit extended family and friends included postering along the way and interviews with rural newspapers and community radio stations. These sorts of things happened continuously, beneath the level of the official and much more conventional publicity strategies coordinated by the project office.

When Cate Jones was hired, she brought her extensive fundraising experience to the table and quickly established an approach to working with foundations and individuals, setting up community events and ensuring consistent press coverage. Cate was inventive when it came to matching particular aspects of the project with particular funding sources: for example, she approached a craftworker fund for $3,000 to support the handstamping of the donor tiles, and she approached engineering

Anne-Marie Edward

Anne-Marie. Notre petit rayon de soleil, continue de nous éclairer dans notre demarche pour l'égalité entre les hommes et les femmes. Ta famille, Suzanne, Jim et Jimmie.

Anne-Marie, our little sunshine, keep lighting our path in our quest for equality between men and women. Your family, Suzanne, Jim and Jimmie.

firms to support construction costs. Her knowledge and expertise also enhanced other committee members' understanding of government and corporate structures and organization.

The group also benefited from members' media experience. Chris was a media technician; ironically, it was her experience with the media that had led her to reject its ephemeral qualities and choose to work on a project that promised a greater degree of permanence. It was her acquaintance with Maya Lin's *Vietnam Memorial* and Judy Chicago's *Dinner Party* that helped her conceive of this project. Even in the early stages, it seems, there was an awareness that documenting this process had to be part of the project. Video footage and photographs were taken during the first few years of the project; records, letters, minutes and other objects were kept throughout; and beginning in 1994, a documentary crew came in to film various aspects of the project. In addition, group members, particularly Suzanne Laplante-Edward (victim Anne-Marie's mother), made themselves available for speaking engagements, presentations, events and interviews, many of which involved the press. Finally, the formal involvement of Lianne Payne (as coordinator in the final years of the project), the national jury of artists and artist Beth Alber resulted in a deeper experience and understanding of artistic sensibilities, the ongoing urban public art debate and the art world that supported the project.

Thus it would seem that an orientation towards feminist process and the understanding from the beginning that this project was intentionally inclusive and likely to be a long-term endeavour characterized by unforeseen consequences ("we just didn't know what they would be") indicate that the group knew it was making history, quite literally. This

Nathalie Croteau

Partie trop tôt avec projets, espoirs et tant de richesses de vie encore à partager. Fleur éternelle, tu occupes maintenant la plus belle place au jardin de nos coeurs. Elise, Fernand, Isabelle Croteau.

Gone too soon with your projects, your hopes, and so much wealth to still share. Eternal flower, you now have the most beautiful place in the garden of our hearts. Elise, Fernand, Isabelle Croteau.

knowledge creates its own kind of inspiration and perhaps enabled group members to frame contemporary difficulties in a way that made them important to bear. It also encouraged among group members the quality of tenacity, a belief in agency and a belief in the value of collective action. As one advocate wrote in a personal note:

> Meetings would really bog down sometimes, especially during the first three or four years, over ways of doing things – whether it was the details of an event, or the way a proposal was written – and a kind of perfectionism would take hold and nothing was quite good enough to let go of and proceed. And I would say something like this, "Let's not perfect every detail of the project. Better we keep moving forward. Women's groups have an undermining tendency towards perfectionism, but what is perfect? There's no such thing (except in nature). It's important that we continue to direct our energy outward and accomplish our goal."

In terms of the boulder project, perhaps the single most helpful attribute was the intimacy with which the organizers understood how things work in the Downtown Eastside. The compassion with which Don Larson spoke about the challenges of overcoming poverty ("everyday, everyday, everyday") reflected the fact that he and the others did not require attendance at meetings and instead focused on points of contact to register support. There was an appreciation of the fact that if such a monument were to be installed, it could not come out of a long process

Maud Haviernick

En mémoire de toi petite soeur, grande fille, belle femme, tout comme ces lettres sur cette pierre, ton visage est à jamais gravé dans nos coeurs. Jamais nous ne t'oublierons, jamais ton sourire, jamais ta joie, jamais toi ... Maud. Ta famille qui t'aime.

In your memory, little sister, big girl, beautiful woman. As these letters are engraved on this stone, your face will be forever engraved in our hearts. We will never forget you, never forget your smile, never forget your joy, never forget you ... Maud. Your family that loves you.

like the one that resulted in the installation of *Marker of Change*. There was neither the money nor the will to mount such a campaign. It is possible, of course, to see the idea of the boulder as opportunistic; some do. But equally, it is possible to see it as a gesture that allowed the Downtown Eastside to be generous when it came time to dedicating *Marker of Change* because "things had [already] been done with the proper respect" with the boulder – just as around the table, though one person may be offered the larger cup of tea, it is the elder who is offered the first one.

Two Aboriginal people played key roles in the boulder project; each individual has considerable standing in the Downtown Eastside. When people introduced themselves and claimed a voice of authority at residents' meetings, people seemed to use the following as self-identifiers:

- claims of work – industries, locations, and physical site details
- residence – where (particular streets/hotels) and length of time
- childhood memories of the Downtown Eastside
- periods of homelessness/foster care/surviving violence
- addiction history/corporeal signs (such as needle marks)

Hard lives count in the Downtown Eastside. Both Aboriginal organizers of the boulder, Fred Arrance and Kelly White, have lived hard lives, and both have extensive family connections in the neighbourhood. Fred has a hook for a left arm and refers to years lost to addiction, has lost at

Michèle Richard

Michèle Richard
Une mélodie, une fleur, une étoile qui scintillera toujours pour nous.
Thérèse Martin Richard, Manon Richard.

Michèle Richard, a melody, a flower, a star that still sheds her light on us.
Thérèse Martin Richard, Manon Richard.

least 14 female members of his own family to violence, and wears his years. Kelly is a relative of one of the missing women, and a friend of several other missing and murdered women. She has repeatedly gone before both the Aboriginal community and the press to bring attention to these cases. She and Fred were, by Fred's account, two of the originators of the Valentine's Day marches. It was Kelly who told the Park Board that the boulder would promote healing, and that it would provide a place for not just Aboriginals to mourn. Kelly and Fred's skills, then, are largely skills of being "of" the neighbourhood – holding legitimate membership, and speaking from the perspective of personal experience.

For his part, Don Larson places himself slightly outside the neighbourhood: "I don't really want to be in the neighbourhood any more – lots of bad vibes now, with the drug use. Every day for twenty years I've been there, but I only lived there for three years." But what Don knows is the world of city parks. He was instrumental in creating CRAB Park, and strategically negotiated a sophisticated involvement of the media at key points; he has since gone on, with the CRAB-Water for Life Society, to create Wendy Poole Park and to work with a new green space near Vancouver General Hospital. He sees his skills as practical: "I'll see things through. I don't like meetings ... But I see something to do, and there's always going to be people saying 'I don't like it,' so, you just have to do it. You know, some opposition or slow democratic process, so you just got to do it." In addition, Don knows a bit about money and has the confidence and stubbornness to "fight City Hall." He claims to have grown up middle class, and his comments about CRAB Park and the

Sonia Pelletier

Sonia Pelletier, tu auras été, durant ta courte vie sur cette planète, notre lumière. Là-haut, continue de penser à nous, comme nous pensons tous à toi. Tes soeurs: Denise, Micheline, Francine, Suzanne, Monique. Tes frères: Normand, Marius. Tes parents: Louise et Ulric Pelletier.

Sonia Pelletier, you have been, during your short life on this planet, our light. Up there, continue to think of us, as we think of you. Your sisters: Denise, Micheline, Francine, Suzanne, Monique. Your brothers: Normand, Marius. Your parents: Louis and Ulric Pelletier.

future central waterfront development indicate a kind of political savvy as well:

> But yeah, the core development should be mixed income housing, a good chunk of core-needy social housing. Not a ghetto, but a mixed development, daycare in there. Make a safe community, a stable safe community, a new community, a waterfront community ... And most of it not high-rise, but medium-rise, lowrise, like 6-7 stories and under ... Well, I'm not against tourism, I'm telling them, if you mix poor people, who you're either kicking out of their housing or [who are] increasingly homeless, who you kick out of their only safe house and ... you're going to be mixing those people with tourists and conventioneers, do you really think that is going to work? Don't you understand that safety is the primary concern of a convention person or tourist?

This combination of skills and opportunity – a legitimate Downtown Eastside membership and voice, a proven track record of success (CRAB Park), an awareness of and ability to court the media, a hard-headed understanding of municipal government processes and time to spend bird-dogging the details (in letters, by phone and in person) – were all relevant in the case of the boulder. And, of course, the relatively modest cost of the boulder meant that it could be commissioned, approved,

Anne-Marie Lemay

Anne-Marie, toi qui as illuminé notre vie, qui nous as manifesté tolérance et compassion; de cet ailleurs, indiques-nous maintenant le chemin de la paix et de l'amour. Isabel, Michelle, et Pierre Lemay

Anne-Marie, you that brightened up our lives, that showed us tolerance and compassion, from this other place, help us now find the way to peace and love. Isabel, Michel, and Pierre Lemay.

designed, engraved and installed within four months, fully paid for by donations from the Larson family.

The street smarts brought to bear for the pole are a bit more complex, not least because witnesses informally claim that hundreds of people were involved one way or another with the project. By the time the plan was set in place for a pole at Oppenheimer, it was being funded through an account held by the City-funded Carnegie Community Centre. This meant that the original overseeing of the project was done by Carnegie – both a much trusted and much *mis*trusted local agency (precisely because it does control/coordinate so much activity and money in the neighbourhood). Indeed, though Oppenheimer Park "is maintained by the Vancouver Parks Board, it is staffed and operated by the Social Planning Arm of the City of Vancouver" in close coordination with the Carnegie Community Centre.[6] Funds for the pole came out of a general revenue pool. There were two outreach officers of Carnegie at the time on site at the park – Steve Johnson (a member of the Nisga'a First Nation) and Kathy Bentall (a European-Canadian woman). There are mixed reports about how facilitative they were perceived to be by the Downtown Eastside residents, though this is common for activists connected to Carnegie.

Nevertheless, it seems that Steve's leadership position made it more likely that his request for a donated pole would be met. To some degree at least, it was the tacit understanding that it was another native person

6 *Old Vancouver* booklet, undated. Copy in possession of the author.

asking for help that brought Dick Baker to work on the pole and help coordinate all the support activities (food, shelter and transportation) that allowed the carvers to live in the area while they worked. Thus, though this pole was carved for "those who have died and those who continue surviving" in the meanest conditions, it required those who were steadily employed to make it possible for the work to be done. Dick provided not only the artistic guidance but also, literally, life-support. He could do this only because he had, first, a lifetime of experience as an artist and, second, a full-time, well-paying job on the docks in North Vancouver.

Of course, Carnegie also provided some logistical support – the buildings to hold the log and lock up the tools, a kitchen and bathroom, and publicity within the City community. All of these speak to the skills and means involved in more traditional community organizing – the provision of office space, publicity, submissions to committees, the writing of monitoring reports and the arrangement of details and materials for the pole-raising day. By Steve's account, he and his coworker and Carnegie did more: they openly solicited community involvement (via word of mouth and *Carnegie Newsletter* pieces). Making the project possible and welcoming was thus no small feat; it was the involvement of such people that made the pole *of* the Downtown Eastside.

In that sense, Steve's understanding of homelessness and what he learned during the project about how to reach people constitute a kind of skill. By his own account, he did not want a particular *kind* of involvement but, rather, a *regular* involvement, and a decrease in people's isolation and alienation. To achieve this goal required a change in attitude (and action) towards addiction, given its prevalence in the Downtown Eastside. The project coordinators' ability to meet people as they presented themselves, and to include them, seems to have been fundamental to the creation of the pole.

There is also a phenomenal concentration of artists residing and working in the neighbourhood of the Downtown Eastside, and a tremendous celebration of art. Some of these artists work in studios, whether conventional or unusual ("I've seen studios on fire escapes and tops of fridges, paintings on Venetian blinds"). Some work with physical materials (carving, painting or drawing on canvas, jewellery or furniture design), some with digital or electronic media (film, photography, video) and others with performance (opera, plays, all kinds of music, dance and street theatre). The back lanes of the area are full of graffiti, murals and poetry; tiny gardens are cultivated even from the soil in old brick walls. There are more formalized activities as well – weekly writing, poetry and storytelling groups, a collective film group, a recent community play that celebrated the one-hundredth anniversary of the Carnegie Community Centre, and ongoing choirs and bands and artist-in-residence programmes. Art in the Downtown Eastside is not merely an aesthetic activity: "It's a healing thing to do, a way of letting out some of the insanity, a way of connecting and healing." Collectively, this sort of embrace and abundance of art and the phenomenal skills demonstrated by the Downtown Eastside residents constituted a huge resource for the pole project.

Proposition: A Politics of Visibility

These elements (the sites, the designs, the varied skill sets, and the need for continuous support and mobilization of the community around the issue of violence) show that in order to break through the screen of denial that sought to cast violence as a personal and gender-neutral crime, advocates had to work creatively across all three domains of public space – the imaginative, the discursive and the physical. While they demonstrated common orientations and tactics, their collective actions suggest another pattern, a set of attributes for what can be understood as a "politics of visibility."

A politics of visibility emerges in part when we recognize that an aspect of what is private must become less so when individual experiences must be collected, expressed and deliberately presented to the body politic. In this sense, the personal is political, because what is cast as private, and how it is maintained as private, has a great deal to do with reinforcing hegemonies and taboos. When grief has been caused by violence, particularly violence not contained by a narrative of war, there is a clash between personal and social responses to grief. A massacre of female students, an apparently systematic murdering of Downtown Eastside residents, and a relentless, often fatal social exclusion practised against residents of a community for most of a century – these are reprehensible social acts. To respond to them with only personal expressions of grief does not adequately acknowledge the magnitude of their social effects.

When private matters become public, crucial considerations include where, how and for how long such matters occupy time and space. Episodic incursions in time and space can be policed, tamed and disciplined. But embodied, repetitive experience demands a doxic and ontologically secure functioning in the world; indeed, spatiality figures greatly in struggles to control the imagination in areas of privacy, inclusion and acceptable social norms. Permanence has thus been used to morally instruct and foreclose debate by all manner of hegemonic social forces. Curiously, when permanence is deployed by hegemonic forces (as noted in the discussion of conventional monuments), it often becomes spectacle. When, however, those outside the hegemony use permanence as a conscious tactic, something beyond spectacle can occur. Instead of a reinforcement of status quo relations, which is then tacitly absorbed by the public, in the case of non-hegemonic uses of permanence there is a persistent erosion of those relations. In other words, when a message that is counter-hegemonic consistently occupies time/space, it gains a sense of continuousness akin to hegemonic messages. It slowly accretes qualities of a (new) standard. *As long as it is in tension* with current events and its environment, as long as other monuments nearby continue to be traditional and, importantly, as long as the monument evokes curiosity and reflection in those who encounter it (Uhrmacher and Tinkler 2008; Harjes 2005), a counter-hegemonic monument (or counter monument) retains the power to surprise and will fade into invisibility. It seems to work instead by continuously eroding taken-for-granted understandings in favour of the contemplation of alternatives. That is, the theatre or stage of the monument becomes a node in a landscape of memory *deliberately* cast in tension with hegemonic monuments.

A politics of visibility necessarily involves several kinds of time:

> The city is temporal. It has been there before, it is now, and it will become. It is a palimpsest of buildings, memories, ideas and traditions (historical time). It also moves at different paces and rhythms, at once fast and slow, braking and accelerating (differential time). And it repeats certain cycles, such as those of seasons, commuting, shopping patterns or the sequences of traffic lights (cyclical time). It is the time of individual perceptions,

The CRAB Park Boulder was accompanied a few years later, and a few feet away, by a City of Vancouver historical marker that explains how these shores were once known as "Luck Lucky." | *Photo: Greg Ehlers*

> from the stillness of boredom and the fast pace of excitement (personal time). And of course, it is the time of the future, of the city yet to come. (Miles, Hall and Borden 2000, 5)

Monuments, of course, work across all these dimensions of time precisely because they are permanent. Like the ball-throwing machine discussed above, monuments speak their message over and over again. When these monument sites are used both ritualistically and on an ad hoc basis, almost as permanent public representatives of more imagined communities, they not only admonish current social relations (rather than past ones, as in the case of the *Mahnmale* monuments) but also reinforce possible ones: "Imagine a world where the war on women is over;" "Imagine a world where poverty is not a crime;" "Imagine feeling safe. All the time." In both respects, this is new territory for monuments.

A politics of visibility also forces experiences of random encounter. In this way, counter-hegemonic monuments can facilitate more inclusive norms for civil society, and thus contribute to generating more emancipatory individual *and* social identities. How one lives in a city is neither entirely unconstrained nor entirely scripted; encountering the Other at unexpected moments finally somewhat equalizes the experience of the marginalized person with that of the (even unintentionally and unconsciously) socially dominant person. Such experiences of hybridity permit each participant to be changed by, and in turn change, the social expectations, performances and outcomes of such an encounter. Any of the encounters at the sites mentioned here – the smudging at *Marker of Change,* the family ceremony at the boulder, the closing circles of the Valentine's Day marches at the pole – are experiences saturated with hybrid exchanges, across gender, race, class, age and ability.

A politics of visibility invites a conscious re-corporealization of public space. *Marker of Change* references bodies explicitly – the size and dimension of the benches akin to "coffins for an average human female," the gashes, the names of thousands and, encircling the benches on tiles, the intimate words of those who chose to speak out. The monument makes space to sit, lie down, have conversations and leave messages, money and food. The pole was fashioned by scores of hands and was raised by hundreds more. As if it is a respected elder, it is brought news in the form of notes, named rocks, candles, tobacco and smudging ceremonies. The boulder, also a site of offerings, is a collective gravestone – a cemetery space, right beside a tourist-pretty walkway. Such disruptions shatter the transparency of abstract spaces, they remove the filter of media, the smoothed narrative, and re-identify tactility as the source of membership and identity.

A politics of visibility alters the norms of commemorative time. These monuments, to use the Swahili understanding, are raised in *sasha* time – while the dead live on in the memories of the living. They are not, according to that schema, to be exhortations to moral order, but instead are more simply expressions of strong emotion. But by deliberately inverting monumental norms – they are not flawless, they do not pretend to aspire to aesthetic distance, they are not heroically proportioned, their inscriptions are not vapid, they are not placed in ceremonial spaces – these monuments have taken on the characteristics of *zamani* monuments, in the sense that they highlight social/political messages. These monuments studied do not simply commemorate events, they *anticipate* them.

Finally, a politics of visibility works consciously and carefully with specificity. In the cases presented here, this specificity centred on naming. Deliberate silence and the presence of absence are features in the textual presentations of all three monuments. *Marker of Change* has left seven of its fourteen benches uninscribed, in reference to the silencing and often unspeakable danger that is part of the story of male violence against women. The boulder quite deliberately leaves names out: partly because not all the names were known; partly because, tragically, there simply wasn't enough room for them all; and partly because there are very particular traditions of permission and respect in First Nations culture surrounding the naming of the dead.

The pole's use of text is slightly more complicated. Because there are conflicting opinions about the merits of writing or specifying in an oral tradition, it is hard to know what weight ought to be given to those small pieces of explanatory text that do exist about the Oppenheimer Park pole. From a literary perspective, all the texts are quite incomplete – that is, they do not correspond very closely to the thing they are describing (for example, several of the figures on the pole are not named in the descriptive texts). There is the difficulty of reading the pole symbolically among its viewers – even among the myriad First Nations viewers themselves. Perhaps the human figures are considered at some level "readable." But even if we do not focus as intently on the texts here as we do on the texts of the other monuments, there are important things to note about the words used.

Specifically, though all three monument texts reference the past, the pole texts heavily emphasize the present. Unlike in *Marker of Change* ("we work for a better world"), there is no reference in the pole texts to agency relating to the future. Rather, the texts honour what it takes to exist ("to those who *have survived* and [who] *continue to live* in the neighbourhood"; my emphasis). There is also reference to violence in a broader sense than is indicated in either of the other monuments – here the violence of poverty, of disease and drugs and of systemic as well as personal inaction are highlighted. Given the general ambivalence about texts where First Nations oral traditions are concerned, it is unusual that the two programmes, the poem and the plaque would all have these characteristics in common, especially given that they were not written by the same authors.

So what constitutes a politics of visibility for counter-hegemonic monuments? First, in terms of a mechanistic analysis, the constellation of factors highlighted above seems necessary to achieve visibility. Second, there must be a sophisticated *spatialized* sense of how memory and permanence can be used tactically. This includes an understanding of the legacies of evictions in specific locations, and how these leave traces readable in the gradations of contemporary continuums of publicness. An awareness that hegemony works not by power *over* but by power *with the consent of* means that advocates of counter-hegemonic monuments need to understand that their memory crafting is problematic for hegemonic forces trying to manufacture a consensus that certain people, places or versions of inhabitance of the city are disposable. They must understand that various efforts will be deployed – to deny, erase or bury – memories that disrupt different representations of the "proper" future for a place. Advocates for the "disposable" (people and places), then, have to be bold – imaginatively, discursively and physically – and consciously work in tension with existing memorial and abstract landscapes by selectively incorporating some monumental norms while simultaneously inverting others. The work is not easy, especially when organizing must be done under conditions of scarcity and assault. Instead, their work is a social gamble that art will resonate beyond the eye to the heart, and that telling the truth in a public place matters. But if such a monument is done well, as Haida artist Bill Reid once offered, "it may be visited by grace. And if it is visited by grace, it has the power of speaking for a long time."

References

Amnesty International Canada. 2004. Stolen sisters: Discrimination and violence against indigenous women in Canada. Report, October 4.

Anderson, B. 1983. *Imagined Communities: Reflections on the Origins and Spread of Nationalism.* London: Verso.

Anderson, K. 1991. Chinatown re-oriented, 1970-1980. In K. Anderson, ed., *Vancouver's Chinatown: Racial Discourse in Canada, 1875-1980,* 211-44. Montreal and Kingston: McGill-Queen's University Press.

Appadurai, A. 1990. Disjuncture and difference in the global culture economy. *Public Culture* 2(2): 1-24.

Bhabha, H.K. 1990. What is a nation? In H.K. Bhabha, ed., *Nation and Narration,* 8-32. London: Routledge.

Blomley, N. 1997. Property, pluralism and the gentrification frontier. *Canadian Journal of Law and Society* 12(2): 187-218.

–. 2004. *Unsettling the City: Urban Land and the Politics of Property.* London: Routledge.

Bouchoutrouch, M. 1993. Facing the backlash. *Kinesis* (September): 4.

Bourdieu, P. 1977. *Outline of a Theory of Practice.* Trans. R. Nice. Cambridge, UK: Cambridge University Press.

Boyer, M.C. 1992. Cities for sale: Merchandising history at South Street Seaport. In M. Sorkin, ed., *Variations on a Theme Park: The New American City and the End of Public Space,* 181-204. New York: Noonday Press.

–. 1994. *The City of Collective Memory: Its Historical Imagery and Architectural Entertainments.* Cambridge, MA: MIT Press.

Bradley, M. 1995. *Reframing the Montreal Massacre: A Media Interrogation* [film]. Toronto: Full Frame Film.

Bula, F. 1992. Time for love, not blame. *Vancouver Sun,* December 9, A17.

Burk, A. 2005. In sight; out of view: A tale of three monuments. *Antipode* 38(1): 41-58.

–. 2006. Beneath and before: Continuums of publicness in public art. *Social and Cultural Geography* 7(6): 949-64.

Byfield, T. 1993. Statue [sic] represents sexism and hatred at its worst. *Calgary Sun,* August 1, C2.

Cameron, Sandy. 2000. The Oppenheimer Park Totem Pole [poem]. In S. Cameron, *Sparks from the Fire,* 70. Vancouver: Lazara.

Cameron, Stevie. 2007. *The Pickton File.* Toronto: Knopf.
Carnegie Newsletters. 1986-2003, biweekly. Vancouver: Carnegie Community Centre.
Carson, B. 2002. *Fix: The Story of an Addicted City* [film]. Nettie Wild, dir. Vancouver: CanadaWild Productions.
Chivallon, C. 2001. Bristol and the eruption of memory: Making the slave-trading past visible. *Social and Cultural Geography* 2(3): 347-63.
Civic committee denies funding for memorial projects. 1996. *Georgia Straight.* July 11-18, 45.
Coppard, J. 1999. *Vancouver Courier,* December 1, 21.
Cresswell, T. 1996. *In Place/Out of Place.* Minneapolis: University of Minnesota Press.
–. 1998. Night discourse: Producing/consuming meaning on the street. In N. Fyfe, ed., *Images of the Street: Planning, Identity and Control in Public Space,* 268-79. London: Routledge.
Cruikshank, J. 1998. *The Social Life of Stories: Narrative and Knowledge in the Yukon Territory.* Lincoln: University of Nebraska Press.
Culbert, L., L. Kines and K. Bolan. 2001. Killer could be getting smarter. *Vancouver Sun,* November 23, A10.
Cultural Memory Group (C. Bold, S. Castaldi, R. Knowles, J. McConnell and L. Schincariol). 2006. *Remembering Women Murdered by Men.* Toronto: Sumach Press.
Czaplicka, J. 1997. Stones set upright in the winds of controversy: An Austrian monument against war and fascism. In S. Denham, I. Kacandes and J. Petropoulus, eds., *A User's Guide to German Cultural Studies,* 257-386. Ann Arbor: University of Michigan Press.
de Oliver, M. 1996. Historical preservation and identity: The Alamo and the production of a consumer landscape. *Antipode* 28(1): 1-23.
Deutsche, R. 1986. Krzystof Wodiczko's *Homeless Projection* and the site of urban "revitalization." *October* 38 (Fall): 63-98.
–. 1998. Uneven development: Public art in New York City. *October* 47 (Winter): 3-52.
Deutsche, R., and C. Ryan. 1984. The fine art of gentrification. *October* 31 (Winter): 91-111.
Didion, J. 2003. Fixed opinions, or the hinge of history. *New York Review of Books,* January 16, 54-59.
Drake, D. 1993. No violence against women: A message that should be cast in stone. *Vancouver Sun,* August 5, A10.
Dunphy, M. 1992. Women seek public support for monument. *Georgia Straight,* July 31-August 7, 4.
Duthie, K. 1993. A monumental work of hatred. *Vancouver Sun,* July 28.
Dwyer, O. 2000. Interpreting the Civil Rights movement: Place, memory and conflict. *The Professional Geographer* 52(4): 660-71.
Fitterman, L. 1999. Cops on Lepine's list: Names of six female officers found on Polytechnique killer. *Montreal Gazette,* March 10, A3.
Foucault, M. 1980. *Power/Knowledge: Selected Interviews and Other Writings 1972-1977.* New York: Pantheon.
Funkenstein, A. 1989. Collective memory and historical consciousness. *History and Memory* 1(1): 5-26.
Gale, H. 1994. Not collective guilt, but collective responsibility. *Globe and Mail,* November 23, A24.
Giddens, A. 1977. *Studies in Social and Political Theory.* London: Hutchinson.
Gill, A. 2001. East side in the eye of the beholder. *Globe and Mail,* June 16, R4.
Gitlin, Todd. 2002. *This Morning,* CBC Radio 1, September 6.

Goffman, E. 1963. *Behavior in Public Places: Notes on the Social Organization of Gatherings.* New York: Free Press of Glencoe.

Gormley, A. 1998. *Making an Angel.* London: Booth-Clibborn Editions.

Gramsci, A. 1971. *Selections from the Prison Notebooks,* edited and translated by Q. Hoare and G. Nowell Smith. London: Lawrence and Wishart.

Grove, N. 2000. Vancouver's Shadows [poem]. In B. Constantine, ed., *A Hurricane in the Basement and other Vancouver Experiences,* 16. Vancouver: City of Vancouver Millennium Collection.

Halbwachs, M. 1952/1992. *On Collective Memory.* Chicago: University of Chicago Press.

Hall, S. 1995. New cultures for old. In D. Massey and P. Jess, eds., *A Place in the World? Places, Cultures and Globalization,* 175-213. Oxford: Open University Press.

Hallendy, N. 2000. *Inuksuit: Silent Messengers of the Arctic.* Vancouver: Douglas and McIntyre.

Haraway, D. *Simians, Cyborgs, and Women: The Reinvention of Nature.* London: Free Association Books.

Harjes, K. 2005. Stumbling stones: Holocaust memorials, national identity, and democratic inclusion in Berlin. *German Politics and Society* 74(23): 138-51.

Harvey, D. 1979. Monument and myth. *Annals of the Association of American Geographers* 69(3): 362-81.

Hayden, D. 1995. *The Power of Place: Urban Landscapes as Public History.* Cambridge, MA: MIT Press.

Heffernan, M., and C. Medlicott. 2000. The first American madonnas: Gender, race and the commemoration of the American frontier, 1890-1930. Paper presented at the American Association of Geographers meeting, Pittsburgh, PA, April 4-8.

Hobsbawm, E. 1995. Foreword. In D. Ades et al., eds., *Art and Power; Europe under the Dictators 1930-45,* 11-15. London: Thames and Hudson.

Hume, S. 1993. Dispatches from the front line of the war on women. *Vancouver Sun,* September 24.

Jacobs, J.M. 1998. Staging difference: Aestheticization and the politics of difference in contemporary cities. In R. Fincher and J.M. Jacobs, eds., *Cities of Difference,* 252-78. New York: Guilford.

Jang, W. 1993. These words speak of hate not honor. *Vancouver Sun,* December 10.

Jay, M. 1992. Scopic regimes of modernity. In S. Lash and J. Lash, eds., *Modernity and Identity,* 178-95. Oxford: Blackwell.

Jensen, V. 1992. *Totem Pole Carving.* Vancouver: Douglas and McIntyre.

Johnston, R.H., D. Gregory and D.M. Smith, eds. 1994. *Dictionary of Human Geography,* 3rd ed. Oxford: Blackwell.

Judt, T. 1998. A la recherche du temps perdu. *New York Review of Books,* December 3, 51-58.

Kines, L., L. Culbert and K. Bolan. 2001. B.C. slow to adopt lessons of Bernardo. *Vancouver Sun,* September 26, A1.

Klein, N. 1997. *The History of Forgetting: Los Angeles and the Erasure of Memory.* New York: Verso.

Komar, V., and A. Melamid. 1994. What is to be done with monumental propaganda? *New Formations* 22 (Spring): 1-3.

Kovanic, G., and G. Johnson. 2000. *Flipping the World: Drugs through a Blue Lens* [film]. Moira Simpson, dir. Montreal: National Film Board.

Kovanic, G., and G. McCrea. 1999. *Through a Blue Lens* [film]. Veronica Alice Mannix, dir. Montreal: Odd Squad Productions Society and National Film Board.

Lautens, T. 1993. The feminine mistake. *Vancouver Sun*, July 22, A15.

Lefebvre, H. 1991. *The Production of Space*. Oxford: Blackwell.

Lett, R. 1993. Mockery of people in pain. *Vancouver Province*, August 10, A24.

Levinson, S. 1998. *Written in Stone: Public Monuments in Changing Societies*. Durham, NC: Duke University Press.

Loewen, J. 1999. *Lies Across America: What Our Historic Sites Get Wrong*. New York: The New Press.

Lowman, J. 2000. Violence and the outlaw status of (street) prostitution in Canada. *Violence Against Women* 6(9): 987-1011.

Mander, J. 1978. *Four Arguments for the Elimination of Television*. New York: Morrow.

Massey, D. 1994. A global sense of place. In *Space, Place and Gender*, 146-56. Cambridge, UK: Polity Press.

McCann, E. 1999. Race, protest, and public space: Contextualizing Lefebvre in the U.S. city. *Antipode* 31(2): 163-84.

Meghill, A. 1998. History, memory, identity. *History of the Human Sciences* 11(3): 37-62.

Michalski, Sergiuz. 1998. *Public Monuments: Art in Political Bondage 1870-1997*. London: Reaktion Books.

Miles, M., T. Hall and I. Borden, eds. 2000. *The City Cultures Reader*. London: Routledge.

Mitchell, D. 1995. The end of public space? People's Park, definitions of the public, and democracy. *Annals of the Association of American Geographers* 85(1): 108-33.

–. 1996a. Introduction: Public space and the city. *Urban Geography* 17(2): 127-31.

–. 1996b. Political violence, order and the legal construction of public space: Power and the public forum doctrine. *Urban Geography* 17(2): 152-78.

Monk, J. 1992. Gender in the landscape: Expressions of power and meaning. In K. Anderson and F. Gale, eds., *Inventing Places: Studies in Cultural Geography*, 123-34. Melbourne: Longman.

Morgan, S. and P. Millar. 1998. *Marker of Change: The Story of the Women's Monument* [film]. Moira Simpson, dir. Victoria, BC: Moving Images Distribution.

Nelson, J. 2000. The space of Africville: Creating, regulating and remembering the urban slum. *Canadian Journal of Law and Society* 15(2): 163-86.

Novick, P. 2000. *The Holocaust in American Life*. New York: Houghton Mifflin Harcourt.

Orwell, G. 1949. *1984*. London: Seeker and Warburg.

Osborn, B. 1989. *East of Main*. Vancouver: Arsenal Pulp Press.

–. 1995. *Lonesome Monsters*. Vancouver: Anvil Press.

Osborne, B. 2001. Landscapes, memory, monuments, and commemoration: Putting identity in its place. Commissioned by the Department of Canadian Heritage for the Ethnocultural, Racial, Religious, and Linguistic Diversity and Identity Seminar, Halifax, NS, November 1-2, 1-38.

Our Nation's Slum: Time to Fix It. 2009. *Globe and Mail*, February-March. http://www.globeandmail.com/thefix.

Park memorializes missing women. 2000. *Vancouver Sun*, March 20, B1.

Peters, E. 2002. You can make a place for it. Presentation at Association of American Geographers Conference, Los Angeles, CA, March 19-23.

Phillips, K., and C. Jones. 1993. Build it and change will begin to come. *Vancouver Sun*, July 1.

Piercy, M. 1980. The Low Road [poem]. In *The Moon is Always Female*. New York: Knopf.

Rayner, W. 1993. Women's monument wrong vehicle for sorrow. *Vancouver Province,* August 18.
Razack, S. 2000. Gendered racial violence and spatialized justice: The murder of Pamela George. *Canadian Journal of Law and Society* 15(2): 91-130.
Ruddick, S. 1996. Constructing difference in public spaces: Race, class and gender as interlocking systems. *Urban Geography* 17(2): 132-51.
Sasaki, K-I. 2000. For whom is city design? Tactility versus visuality. In M. Miles, T. Hall and I. Borden, eds., *The City Cultures Reader,* 36-44. London: Routledge.
Sennett, R. 1970. *The Uses of Disorder.* New York: Vintage.
Service planned for missing women. 1999. *Vancouver Sun,* May 10, B1.
Sommers, J. 2001. The place of the poor: Poverty, spaces and the politics of representation in downtown Vancouver, 1950-1997. PhD diss., Simon Fraser University.
–. 2002. Living in the shadow of the world class city: Countering marginality in Vancouver's downtown eastside [pamphlet].
Stallybrass, P., and A. White. 1986. Introduction. *Politics and Poetics of Transgression,* 1-15. Ithaca: Cornell University Press.
Sturken, M. 1997. *Tangled Memories: The Vietnam War, the AIDS Epidemic, and the Politics of Remembering.* Berkeley: University of California Press.
Till, K. 1999. Staging the past: Landscape designs, cultural identity and Erinnerungs-politik at Berlin's Neu Wache. *Ecumene* 6(3): 251-83.
Todd, D. 1998. Monuments policy toughened. *Vancouver Sun,* December 11, A3.
Uhrmacher, P., and B. Tinkler. 2008. Engaging learners and the community through the study of monuments. *International Journal of Leadership in Education* 11(3): 225-38.
Walker, Alice. 2001. *The Way Forward Is with a Broken Heart.* New York: Ballantine.
Warner, M. 1996. *Monuments and Maidens.* New York: Vintage.
Whittlesea, B.W.A. 1997. Production, evaluation and preservation of experience: Constructive processing in remembering and performance tasks. In D.L. Medin, ed., *The Psychology of Learning and Motivation,* vol. 37, 211-64. New York: Academic Press.
Wood, D. 1999. Missing. *Elm Street,* November, 96-110.
Woolford, A. 2001. Tainted space: Representations of injection drug use and HIV/AIDS in Vancouver's DTES. *BC Studies* 129 (Spring): 27-52.
Young, J. 1993. *The Texture of Memory: Holocaust Memorials and Meaning.* New Haven and London: Yale University Press.
–. 2000. *At Memory's Edge.* New Haven: Yale University Press.
Younge, G. 2002. Premature adulation. *Guardian Weekly,* July 11-17, 13.

Index

Printed and bound in Canada by Friesens
Set in Korinna and Minion by Artegraphica Design Co. Ltd.
Copy editor: Jillian Shoichet
Proofreader: Stephanie VanderMeulen